C. H. F. Rowell

Locust Neurobiology

A Bibliography, 1871–1991

Birkhäuser Verlag
Basel · Boston · Berlin

Author's address:
Prof. Dr. C.H.F. Rowell
Zoologisches Institut der Universität
Rheinsprung 9
CH-4051 Basel/Switzerland

Deutsche Bibliothek Cataloging-in-Publication Data

Rowell, Charles H. F.: Locust neurobiology : a bibliography, 1871 – 1991 / C. H. F. Rowell. - Basel ; Boston ; Berlin : Birkhäuser, 1992
ISBN 3-7643-2747-2 (Basel ..., mit Diskette DOS-Version) brosch.
ISBN 0-8176-2747-2 (Boston, mit Diskette DOS-Version) brosch.
ISBN 3-7643-2748-0 (Basel ..., mit Diskette Macintosh-Version) brosch.
ISBN 0-8176-2748-0 (Boston, mit Diskette Macintosh-Version) brosch.
NE: HST

Printed from the author's camera-ready manuscript on acid-free paper in Germany

ISBN 3-7643-2748-0
ISBN 0-8176-2748-0
[Macintosh version]

ISBN 3-7643-2747-2
ISBN 0-8176-2747-2
[DOS version]

Contents

FOREWORD

The acridoid grasshoppers in general and the various species of swarm-forming locusts in particular have been among the most favoured subjects of insect physiology, behaviour and ecology for many years. Several factors contribute to this popularity: their abundance in nature and their ease of culture in the laboratory, their relatively large size, and most of all, perhaps, their agricultural importance and the consequent intermittent availability of funds for their study. These factors together have inspired a large amount of experimental work, and this in turn has produced a new and often over-riding reason for working on acridoids - the huge body of available background information and know-how that has built up about these insects.

This state of affairs is well seen in insect neurobiology. Only a restricted number of insect types are commonly used in this discipline, and originally most of them were selected for reasons of convenience and availability: grasshoppers, cockroaches, crickets, flies, bees and moths are the most important. Each of these insects is the subject of the attentions of one or more major groups of neurobiologists, but neurobiological articles on acridoids probably exceed in number those on all other insects combined, at least if articles on the molecular biology of the nervous system of *Drosophila* are excluded.

Paradoxically, this wealth of information has become almost a hindrance to research, for it is becoming increasingly hard to retain an overall view of the whole, and the older contributions are often forgotten. The situation is exacerbated by the fact that neurobiological research using grasshoppers is usually of no direct interest to agriculture, medicine or biotechnology, and is consequently omitted from the large abstracting services and computer data-banks which are financed directly or indirectly by these industries.

It is for these reasons that I have compiled this bibliography, in the hope that it will allow the next generation of researchers a rapid access to the knowledge gained in the past. It is as complete as I could make it. Most of the authors who have published in the past two decades have been circularized and asked for corrections and additions. However, there will inevitably be many mistakes and omissions. This is especially true for articles written in Slavic or Asian languages, as my own reading is limited to Western European languages. I apologize in advance to authors whose works have been overlooked or miscited, and would ask them to send me reprints and their publication lists, in case this bibliography is ever reissued in a revised version.

I would be most grateful if users would send me corrections and additions to the bibliography as they become aware of them. The address is

Prof. C.H.F. Rowell,
Dept. of Zoology, University of Basel,
Rheinsprung 9,
4051 Basel,
Switzerland.

PREFACE

1. The choice of articles to be listed.

The core of this compilation consists of titles of works which make direct reference to the nervous system or the muscles of acridiod grasshoppers, or to their constituent cells and subcellular components. I interpret the adjective "acridoid" widely for these purposes, including among others the families Acrididae, Pamphagidae, Pyrgomorphidae, and Romaleidae, and the subfamily Morabinae. The absence of references to other caeliferan taxa, for example the other eumastacoids and the tetrigids, is due to the dearth of physiological literature on these and similar groups, and not to any taxonomic bias on my part.

In order to restrict the size of the project, I have reluctantly omitted all works dealing with other orthopteroid insects, even the relatively closely related gryllids and tettigoniids. The resultant gap is to some extent filled by the recent publication of the symposium *Cricket behaviour and neurobiology* (eds. F. Huber, T.E. Moore, & W. Loher. Ithaca & London: Cornell University Press. Pp. 565 (1989)) which includes extensive bibliographies. The file on which the text is based was closed in January 1992.

Within the acridiod literature there are however several fringe areas where my selection has been arbitrary and eclectic. One of these is the area of hormones and neuropeptides. I have in general included works on peptide hormones, on the grounds that these are usually the product of neurons, and conversely steroid hormones and juvenile hormone have been omitted. I have not however included works which are concerned only with the interaction of peptide hormones and their target organs, if the latter do not include muscle or nerve cells, and I have included works on the innervation (but not the functioning) of the corpus allatum, which does not produce neuropeptides. Another difficult area is that of behaviour. Here I have included two categories of papers: first, those that advance hypotheses of causation at the neural level or describe movement in terms of muscle activity, and secondly, those describing behaviour patterns that are or have been the subject of detailed neurophysiological research. This latter criterion has caused me to include purely behavioural work on flight, jumping, stridulation and ingestion, for instance, but to ignore similar work on other forms of courtship or the selection of substrates for egg-laying. I have also included a number of titles dealing with the composition of the haemolymph and with physiological saline solutions and other technical matters of direct interest to experimental neurobiologists. Works on gross anatomy are all included, as neurobiologists are continually in need of technical anatomical names.

Another grey area concerns dissertations, "Habilitations-schriften", published abstracts, symposium contributions, and general articles which include acridid examples. In theory I would have preferred to include only those titles from these categories that contain information otherwise not published in refereed journals. Often, however, it is difficult to ascertain whether or not this is the case. A very few authors have indicated to me which Abstracts can be dropped from the list of their publications, and sometimes I have done this on my own initiative, where it is clear that all the information has been superceded by a full publication elsewhere. There remain however many recent abstract titles, the majority of which probably either are already or will be superfluous.

Finally, comprehensiveness implies that all the published works are listed, be they modern or superceded, careful or slipshod, confirmed or disproven, fundamental or trivial. Caveat lector!

2. How to use this book.

The **printed version** of this bibliography takes the traditional form of an alphabetical listing by author and by year. Frequently one knows the name of the author of the work one wishes to cite or reread, and the traditional alphabetical listing is then the most convenient. The order is that produced by sorting the data-base for author, year and journal, in that order of precedence. Each entry also includes key words (note that these have been provided by me, and not by the original author), which often supplement the information contained in the title. More about the keywords can be found below.

The **data-base version** is infinitely superior to the printed one for research purposes. With an appropriate programme it is possible to search the entire bibliography for keywords and produce lists of the articles relevant to any given topic, and then to order them by date or author or journal; or to find all the publications of a given author, and so on. For such purposes the keywords are usually more useful than the words contained in the title. These latter are often not very explicit, and there is the further problem of synonomy, for example "audition", "hearing", "response to sound", "auditory interneuron", "ear" and "tympanum" are closely related concepts but require separate searches of the databank. All would be found by searching for the keyword **aud**. A list of the keywords used and their abbreviations is given below, immediately preceding the bibliography proper, together with a description of the rationale used in allotting them.

The data-base is supplied on diskette in the form of an ASCII text file, in both Macintosh and DOS formats. Fields (of which there are five, given in the order AUTHOR/ YEAR/ TITLE/ PUBLICATION/ KEYWORDS) are separated by tabulator symbols, and entries are separated by paragraph symbols. This format allows the database to be imported directly into virtually all commonly used database, filing and bibliographic programmes. I have used it with great effect with the Microsoft products File, Excel, and Word 4, and with Claris Filemaker, all in their Macintosh versions. As with all computer applications, familiarity with the programme (and its Manual!) is essential if good results are to be obtained, and the faster the computer the better the service. Normal precautions, such as making several reserve copies of the original data disk before attempting to manipulate its contents, should be observed.

I have not split up the bibliographic information, but have left journal title, volume number, and first and last page numbers in one field. Some users may wish to insert tabulator symbols between these values, for example, in order to allow them to be separately manipulated in a bibliographic programme which formats according to the idiosyncracies of different journals. In my own experience, one rarely cites so many references in a given article that it becomes advantageous to use automatic formatting functions, which are therefore often more trouble than they are worth. Should it however be necessary to insert tabulator symbols, this can most easily be done by importing the text file as supplied into a data base programme which permits search and replace within individual fields, and using this function to replace with distinctive arbitrary symbols the colons, dashes and so on which separate the desired quantities. The contents of the data base can then be output as a text document again, and with the search and replace functions of a word processor the symbols replaced by tabulator symbols, and the text edited. Finally, the resultant document can be reimported into a database with the necessary additional new fields.

3. The key words.

The list of key words is given below. In the bibliography they are given in abbreviated form if they are more than 5 or 6 letters long. This abbreviation is designed to be as easily recognizable as possible, while remaining unambiguous when sorted by computer. Where practical, the abbreviation consists of the initial letters of the word: in this case it is written in lower case letters (e.g. **musc** for muscle, **degen** for degeneration, **oc** for ocellus, and so on). Sometimes two words are telescoped together and the division is then indicated by a capital letter,(e.g., **fineStruc** and **patRec** for fine structure and pattern recognition, respectively). Where a simple abbreviation is not practical, I have used an acronym composed of the initial letters of several words, and these are then written in capital letters (e.g. **IN** for interneuron, **MN** for motor neuron, **NSC** for neurosecretory cell). Combinations of these two conventions can easily be recognized as such, as in **ACh** for acetylcholine, **CAll** for corpus allatum. Standard chemical abbreviations have been retained (e.g. **gly, cAMP, GABA, 5HT,** and so on), as have a few time-honoured abbreviations such as **mpt** for mouthparts.

A taxonomic key word has been allotted only if either a) the grasshopper treated in the reference is not a member of the Acrididae (e.g., if it is a Romaleid the abbreviation **rom** would be given) or b) the article includes work on another Order of insects as well (e.g. if work on cockroaches were included, then the abbreviation **dict** (for Dictyoptera) would be given). In the abence of a taxonomic key word it can be assumed that the article deals only with Acrididae in the strict sense.

After a little practice the user will find that he can usually guess the abbreviation of any given keyword, and reference to the printed list will become unnecessary. If the use of a guessed abbreviation produces ambiguity, then it is usually because it has been made too short - for example, searching for **pro** will find references dealing with proctolin (**proct**) proprioception (**propr**) and the prothoracic gland (**prothG**). In the interests of "user friendliness", some keywords are hierarchically redundant. For example, all entries of **CC** and **CAll** (corpus cardiacum and corpus allatum repectively) also include the key word **retroc** (retrocerebral complex). Similarly the term "transmitter substance" (**TS**) is used redundantly in addition to the names of specific chemicals, and the term "sense organ" (**SO**) indicates virtually all types of sensilla, which are further specified by such epithets as **chem**, **hygr**, **mech** and so on (it does not however include photoreceptors, which are subsumed under "retina", **ret**). The user of the computer version of this data base will however probably wish to add further keywords to suit his or her special interests. I have for example grouped all aspects of developmental biology and embryology under the single abbreviation **devel** - other readers may wish to add more detail.

When searching by computer for keywords it is advisable to include the space after the word in the search string. This will obviate finding false positives in the form of letter combinations embedded in some other word. For example, the string **SO** (meaning sense organ), if specified without a terminal space, and assuming that the search function does not differentiate between capital and small letters, will also find the keyword **synapto<u>so</u>me**, though I have tried to reduce such ambiguities to a minimum. For this reason, if a user finds it necessary to add new keywords to an entry, care should be taken to maintain a space between the last keyword and the paragraph symbol.

Table of keywords and their abbreviations

2-DOG (2-deoxyglucose)
5HT (5-hydroxytryptamine)

abd(omen)
ACh (acetylcholine)
ADH (antidiuretic hormone)
aero(dynamics)
AKH (adipokinetic hormone)
allatost(atin)
anat(omy)
ant(enna)
asp(artate)
aud(ition, auditory)
autot(omy of the leg)

batEv(asion)
behav(iour)
blood
BPP (bovine pancreatic polypeptide)
brain
brainStim(ulation)
brun(ner's organ)
bursicon

Ca2 (calcium ions)
CAll (corpus allata)
cAMP (cyclic adenosine monophosphate)
CB (central body)
CC (corpus cardiacum)
CCAP (crustacean cardioactive peptide)
cDNA
cellcult(ure)
cerc(us)
channel (membrane ion channel)
chem(oreception)
CO (chordotonal organ)
col(our)
CPCH (crustacean pigment concentrating hormone)
CPDH (crustacean pigment dispersing hormone)

DD (dorsal diaphragm)
degen(eration)
desens(itisation of ligand receptors)
devel(opment)
DH (diuretic hormone)
diap(ause)
dict(yoptera)
dimorph(ism, sexual)
dipt(era)
DN (descending neuron from brain)
dopa(mine)
DUM (dorsal unpaired medial neuron)

ECM (extracellular matrix)
elect(rical synapse, or electrical inhibition)
enkeph(alin)
enz(yme)
ephem(eroptera)

facNonresp facultative non-responsiveness
FCO (femoral chordotonal organ)
feed(ing)
fineStruc (fine structure)
flight
FMRF(amide)

G-prot(ein)
GABA (g-aminobutyric acid)
genet(ics)
genit(alia, genital tract, gonads)
GF (giant fibre)
glia
glu(tamate)
gly(cine)
groom(ing behaviour)
gryll(idae)
gut

habit(uation)
head
heart
HF (high frequency sound)
hist(amine)
homol(ogy)
hormX (unidentified hormone)
hygr(oreception)

IN (interneuron)
inhib(ition)
innerv(ation)
insP (inositol phosphate)
insul(in)

JH (juvenile hormone)
jump (see also "**kick**")

kick (see also "**jump**")

labil(ity)
LCCP (locust CC peptide)
learn(ing)
leg
lent(ulidae)
lep(idoptera)
lip(id derivative)
local (local interneuron)
LoMIP (locust myoinhibiting peptide)
LoMT (locustamyotropin)
loPyr (locustapyrokinin)
loTach (locustatachykinin)

MB (mushroom bodies)
mech(anoreception)
medN (median nerve)
metab(olism)
MN (motor neuron)
mod(ulation, modulator)
model(ling, mathematical simulation)
mor(abinae)
moult moulting, ecdysis
mpt (mouthparts)
musc(le, musculature)

neck
neuropars(in)
neurophys(in)
NM (neuromuscular junction)
nonspik(ing neuron)
norad(renalin)
NPY (neuropeptide Y)
NSC (neurosecretory cells)

oc(ellus)
oct(opamine)
odon(ata)

olf(action)
OM (optomotor)
opt(ic lobe)
ovip(osition)

pamph(agidae)
patRec (pattern recognition)
pep(tide)
pH
phase (i.e. solitary v. gregarious)
pher(omone)
pneu(moridae)
proct(olin)
propr(ioception)
prothG (prothoracic gland)
pyrg(omorphidae)

recept(or molecule for a ligand)
reflex
regen(eration)
resp(iration)
ret(ina)
retroc(erebral complex)
rev(iew)
rhythm
rom(aleidae)

saline
saliv(ary)
secr(etion of a gland)
SGG (stomatogastric nervous system)
sheath (of nervous system)
SO (sense organ)
spir(acle)
SR (stretch receptor)
strid(ulation)
subg(enual organ)
suboes(ophageal ganglion)
SubP (substance P)
swim(ming)
synaptosome
synon(omy of neuron names)

taur(ine)
tech(nique)
teg(ula)
temp(erature)
tett(igonidae)
therm(oreception)
thor(ax)
tox(icology, toxin)
trach(ea)
TS (transmitter substance)
tymp(anum)

var(iation)
vasop(ressin)
VD (ventral diaphragm)
vibr(ation)
vision
VUM (ventral unpaired medial neurones)

walk(ing)
wind
windh(airs)
wing

BIBLIOGRAPHY

ABISGOLD JD SIMPSON SJ 1988 The effect of dietary protein levels and haemolymph composition on the sensitivity of the maxillary palp chemoreceptors of locust. *J. exp. Biol. 133: 215-229* **feed SO chem mpt blood**

ABRAMS T PEARSON KG 1982 The effect of temperature on identified central neurons that control jumping in the grasshopper. *J. Neurosci. 2: 1538-1553* **aud IN temp thor**

ABUSHAMA FT 1968 Food-plant selection by *Poekilocerus hieroglyphicus* (Klug) (Acrididae: Pyrgomorphinae) and some of the receptors involved. *Proc. R. entomol. Soc. Lond. Ser A 43: 96-104* **feed chem SO mpt pyrg**

ADAM L-J 1968 Verarbeitung akustischer Reize im Insektengehirn. *Naturwiss. 12: 654-655* **aud IN brain**

ADAM L-J 1968 Funktionsbereiche einzelner Hörneurone im Gehirn und Tonbereiche arteigener Laute der Heuschrecke, *Locusta migratoria. Verh. dt. Zool. Ges. 1968: 623-629* **aud IN brain strid**

ADAM L-J 1969 Neurophysiologie des Hörens und Bioakustik einer Feldheuschrecke (*Locusta migratoria*). *Z. vergl. Physiol. 63: 227-289* **aud IN brain**

ADAM L-J 1970 Überlegungen zur Verarbeitung von Nachrichten im Zentralnervensystem von Insekten. Untersuchungen am Hörsystem einer Feldheuschrecke. *Verh. dt. Zool. Ges. 1970: 245-249* **aud IN brain**

ADAM L-J 1977 The oscillating summed action potential of an insect's auditory nerve (*Locusta migratoria*, Acrididae). I. Its original form and time constancy. *Biol. Cybern. 26: 241-247* **aud SO**

ADAM L-J 1977 The oscillating summed action potential of an insect's auditory nerve (*Locusta migratoria*, Acrididae). II. Underlying spike pattern and causes of spike synchronization. *Biol. Cybern. 28: 109-119* **aud SO tymp**

ADAM L-J 1977 Preferential directions of nervous "intensity filters"; auditory neurons in the brain of *Locusta migratoria* (Acrididae). *Zool. Jb. (Physiol.) 81: 250-272* **aud IN brain**

ADAM L-J 1983 Hind femora influence directionality of the locust ear. A possible physical solution for the front/rear ambiguity. *J. comp. Physiol. A 152: 509-516* **aud SO tymp**

ADAM L-J SCHWARTZKOPFF J 1967 Getrennte nervöse Representation für verschiedene Tonbereiche im Protocerebrum von *Locusta migratoria. Z. vergl. Physiol. 54: 246-255* **aud brain SO tymp**

AFIFI AM 1958 On the post-embryonic development of a compound nerve-ganglion in *Locusta migratoria migratorioides* (R. & F.) *Bull. Soc. entomol. Egypte 42: 461-464* **anat devel thor**

AFIFI AM 1959 On a sexual dimorphism in the olfactory lobe of the migratory locust *Locusta migratoria migratorioides* (R. & F.). *Bull. Soc. entomol. Egypte 43: 315-317* **anat ant pher brain dimorph**

AFIFI AM 1960 Ueber die postembryonale Entwicklung des Zentralnervensystems (ZNS) bei der Wanderheuschrecke *Locusta migratoria migratorioides* (R. & F.) (Orthoptera - Acrididae). *Zool. Jb. (Anat.) 78: 1-38* **anat devel**

AGUILAR JS FONSECA MI LUNT GG 1989 Differential effect of ethanol on muscarinic cholinergic binding to rat and locust neural membranes. *Neurochem. Res. 14: 765-770* **TS ACh**

AGUILAR JS LUNT GG 1984 Cholinergic binding sites with muscarinic properties on membranes from the supraesophageal ganglion of the locust (*Schistocerca gregaria*). *Int. Neurochem. 6: 501-507* **brain ACh recept**

AIDLEY DJ 1963 Influence of calcium ions on potassium contracture in an insect leg muscle. *Nature 198: 591-592* **musc ca2**

AIDLEY DJ 1965 The effects of strontium and other divalent cations on potassium contracture in a locust leg muscle. *J. Physiol. (Lond.) 177: 103-111* **musc**

AIDLEY DJ 1965 The effect of calcium ions on potassium contracture in a locust leg muscle. *J. Physiol. (Lond.) 177: 94-102* **musc ca2**

AIDLEY DJ 1985 Muscular contraction. *In: Kerkut GA Gilbert LI (eds.), Comprehensive insect physiology biochemistry and pharmacology, Vol. 5. London: Pergamon Press. 407-437* **musc**

ALBRECHT O 1953 *The anatomy of the migratory locust. London: Athlone Press. 118 pp.* **anat**

ALBRECHT O 1956 The anatomy of the red locust (*Nomadacris septemfasciata* Serville). *Anti-locust Bull. 23. London: HMSO. 9 pp.* **anat**

ALEXANDER R McN BENNET-CLARK HC 1977 Storage of elastic strain energy in muscle and other tissues. *Nature 265: 114-117* **jump musc**

ALICATA P 1962 Muscolatura e sistema nervoso del torace di *Eyprepocnemis plorans* (Charp.) e consideratione sul sistema nervoso cervico-toracico degli insetti. *Arch. Zool. ital. 47: 263-337* **anat thor musc neck**

ALICATA P 1962 Morfologia comparata ed evoluzione della regione del collo negli Ortotteri. *Boll. Zool. 29: 49-162* **anat thor musc neck**

ALLAIS JP GRIPOIS D MORETEAU B RAMADE F 1979 N-acetylation enzymatique de la tryptamine par des homogénats de cerveaux de *Locusta migratoria cinerascens*, avant et après intoxication par le chlordimiform ou le lindane. *Experientia 35: 1357-1359* **metab tox**

ALLUM RC 1973 Surgical interference with the anterior stomatogastric nervous system and its effect on growth and moulting in *Locusta migratoria migratorioides* R. & F. *Dissertation, University of Nottingham, U.K.* **SGG moult**

ALTMAN JS 1975 Changes in the flight motor pattern during the development of the australian plague locust, *Chortoiocetes terminifera. J. comp. Physiol. A 97: 127-142* **flight devel rhythm**

ALTMAN JS 1982 The role of sensory inputs in insect flight motor pattern generation. *Trends in Neurosci. 5: 257-260* **flight SO**

ALTMAN JS 1983 Sensory inputs and the generation of the locust flight motor pattern: from the past to the future. *In: Nachtigall W (ed.), BIONA-report 2 "Insekten Flug"; Akad. Wiss. Mainz. Stuttgart & New York: Gustav Fischer Verlag. 127-136* **flight SO**

ALTMAN JS 1976 The functional organization of the thoracic ganglia in insects. *In: Broughton WB (ed.), Insect neuroacoustics. London: City of London Polytechnic Press. 4-10* **anat MN thor**

ALTMAN JS 1981 Functional organization of insect ganglia. *In: Salánki J (ed.), Neurophysiology of invertebrates. Adv. Physiol. Sci. 23. Oxford: Pergamon Press and Budapest: Akadémiai Kiadó. 537-555* **anat**

ALTMAN JS ANSELMENT E KUTSCH W 1978 Postembryonic development of an insect sensory system: ingrowth of axons from hindwing sense organs in *Locusta migratoria. Proc. R. Soc. Lond. (Biol.) 202: 497-516* **anat teg devel propr SO**

ALTMAN JS BELL EM 1973 A rapid method for the demonstration of nerve cell bodies in invertebrate central nervous systems. *Brain Res. 63: 487-489* **tech**

ALTMAN JS KIEN J 1979 Suboesophageal neurones involved in head movements and feeding in locusts. *Proc. R. Soc. Lond. (Biol.) 205: 209-227* **anat SO suboes saliv windh SO mech feed**

ALTMAN JS KIEN J 1985 The anatomical basis for intersegmental and bilateral coordination in locusts. *In: Bush BMH Clarac F (eds.), Coordination of motor behaviour. Soc. Exp. Biol. Seminar Series 24, Cambridge University Press, Cambridge. 91-120* **anat IN**

ALTMAN JS KIEN J 1987 Functional organization of the suboesophageal ganglion in arthropods. *In: Gupta AP (ed.), Arthropod brain: its evolution, development, structure and function. New York: John Wiley & Sons Inc. 265-322* **anat suboes**

ALTMAN JS KIEN J 1988 A model for decision-making in the insect nervous system. *In: Ali MA (ed.) Nervous systems in invertebrates. New York: Plenum Press. 621-643* **IN brain DN thor abd**

ALTMAN JS KIEN J 1989 New models for motor control. *Neural Computation 1: 173-183* **thor IN walk**

ALTMAN JS SHAW MK TYRER NM 1979 Visualization of synapses of physiologically identified cobalt-filled neurones in the locust. *J. Physiol. (Lond.) 296: 2-3P* **fineStruc anat**

ALTMAN JS SHAW MK TYRER NM 1980 Input synapses on to a locust sensory neuron revealed by cobalt-electron microscopy. *Brain Res. 189: 245-250* **fineStruc**

ALTMAN JS TYRER NM 1974 Insect flight as a system for the study of the development of neuronal connections. *In: Barton Browne L (ed.), Experimental analysis of insect behaviour. Berlin, Heidelberg, New York: Springer Verlag. 159-179* **flight MN SR anat propr devel**

ALTMAN JS TYRER NM 1977 The locust wing hinge stretch receptors. I. Primary sensory neurones with enormous central arborizations. *J. comp. Neurol. 172: 409-430* **anat SR SO thor flight propr**

ALTMAN JS TYRER NM 1977 The locust wing hinge stretch receptors. II. Variation, alternative pathways, and "mistakes" in the central arborizations. *J. comp. Neurol. 172: 431-440* **anat var SR SO thor flight propr**

ALTMAN JS TYRER NM 1980 Filling selected neurones with cobalt through cut axons. *In: Strausfeld NJ Miller TA (eds.), Neuroanatomical techniques. Insect nervous system. Berlin, Heidelberg, New York: Springer Verlag. 373-402* **tech**

ALTNER H AMEISMEIER F 1986 Tubular bodies in dendritic outer segments projecting to second embryonic cuticle from Anlagen of contact chemoreceptors in *Locusta migratoria* L. (Orthoptera: Acrididae) and

Periplaneta americana (L.) (Dictyoptera: Blattidae). *Int. J. Insect Morphol. Embryol. 15: 253-262* **SO ant chem fineStruc dict**

ALTNER H PRILLINGER L 1980 Ultrastructure of invertebrate chemo-, thermo-, and hygroreceptors and its functional significance. *Int. Rev. Cytol. 67: 69-139* **SO ant temp chem hygr fineStruc rev**

ALTNER H ROUTIL C LOFTUS R 1981 The structure of bimodal chemo-, thermo-, and hygroreceptive sensilla on the antenna of *Locusta migratoria. Cell Tiss. Res. 215: 289-308* **SO ant temp chem fineStruc**

AMEISMEIER F 1985 Embryonic development and moulting of the antennal coeloconic no pore- and double-walled wall pore sensilla in *Locusta migratoria* (Insecta, Orthopteroidea). *Zoomorphol. 105: 356-366* **SO fineStruc ant devel moult**

AMEISMEIER F 1987 Ultrastructure of the chemosensitive basiconic single-walled sensilla on the antennae in adults and embryonic stages of *Locusta migratoria* L. (Insecta, Orthoptera). *Cell & Tiss. Res 247: 605-612* **SO fineStruc ant devel chem**

AMEISMEIER F LOFTUS R 1988 Response characteristics of cold cell on the antenna of *Locusta migratoria. J. comp. Physiol. A 163: 507-516* **ant SO therm**

AMMERMÜLLER J 1986 Passive cable properties of locust ocellar L-neurons. *J. comp. Physiol. A 158: 339-344* **oc IN brain**

AMMERMÜLLER J WEILER R 1985 S-neurons and not L-neurons are the source of GABAergic action in the ocellar retina. *J. comp. Physiol. A 157: 779-788* **oc IN GABA TS brain**

AMMERMÜLLER J ZETTLER F 1986 Time- and voltage-dependent currents in locust ocellar L-neurons. *J. comp. Physiol. A 159: 363-376* **oc IN brain**

ANDERSON CR CULL-CANDY SG MILEDI R 1977 Potential-dependent transition temperature of ionic channels induced by glutamate in locust muscle. *Nature 268: 663-665* **NM temp TS glu musc**

ANDERSON H 1978 Postembryonic development of the visual system of the locust, *Schistocerca gregaria.* I. Patterns of growth and developmental interactions in the retina and optic lobes. *J. Embryol. exp. Morphol. 45: 55-83* **opt anat devel**

ANDERSON H 1985 The distribution of mechanosensory hair afferents within the locust central nervous system. *Brain Res. 333: 97-102* **anat windh SO mech**

ANDERSON H 1985 The development of projections and connections from transplanted locust sensory neurons *J. Embryol. exp. Morphol. 85: 207-224* **anat windh mech SO devel**

ANDERSON M COCHRANE DG 1977 Studies on the mid-gut of the desert locust, *Schistocerca gregaria.* I. Morphology and electrophysiology of the muscle coat. *Physiol. Entomol. 247-243* **anat musc gut**

ANDERSON M COCHRANE DG 1978 Studies on the mid-gut of the desert locust, *Schistocerca gregaria.* II. Ultrastructure of the muscle coat and its innervation. *J. Morphol. 156: 257-278* **gut innerv fineStruc musc**

ANDERSON M COCHRANE DG ELDER HY JOSEPHSON RK USHERWOOD PNR 1970 Structure and function of phasic and tonic muscle fibres in locusts and grasshoppers. *J. Physiol. (Lond.) 212: 16-17P* **musc anat**

ANIS NA CLARK RB GRATION KAF USHERWOOD PNR 1981 Influence of agonists on desensitization of glutamate receptors on locust muscle. *J. Physiol. (Lond.) 312: 345-364* **TS glu musc desens recept**

ANIS NA EVANS M USHERWOOD PNR 1979 Plant lectins and desensitization of locust glutamate receptors. *J. Physiol. (Lond.) 291: 47P* **TS glu musc desens recept**

ANONYMOUS 1989 Desert locusts cross the Atlantic from Africa. *ODNRI Newsletter 4, January 1989: 3-4* **flight behav**

ANWYL R USHERWOOD PNR 1974 Voltage-clamp studies of the glutamate response at the insect neuromuscular junction. *J. Physiol. (Lond.) 242: 86-87P* **TS glu NM**

ANWYL R USHERWOOD PNR 1974 Voltage-clamp studies of glutamate synapse. *Nature 252: 591-593* **NM glu TS**

ANWYL R USHERWOOD PNR 1975 The ionic permeability changes caused by the excitatory transmitter at the insect neuromuscular junction. *J. Physiol. (Lond.) 249: 24-25P* **TS glu NM channel**

ANWYL R USHERWOOD PNR 1975 Factors affecting the time course of decay of excitatory post-synaptic currents at a glutamate synapse. *J. Physiol. (Lond.) 254: 46-47P* **TS glu NM channel**

APPLEBAUM SW GADOT M HIRSCH J EL-HADI FA 1991 Allatal stimulation and inhibition in locusts. *In: Menn JJ Kelly TJ Masler EP (eds.), Insect neuropeptides: chemistry, biology and action. Washington DC: American Chemical Society. 450-476* **allatost NSC retroc CAll pep**

ARANYI Z HIRIPI L S-RÓZSA K 1981 Distribution and metabolism of octopamine in the locust brain as related to metamophosis. *In: S-Rózsa K* *(ed.), Neurotransmitters in invertebrates. Budapest: Pergamon Press. Adv. Physiol. Sci. 22: 179-188* **oct brain**

ARBAS EA 1978 Aerial manoeuvring in orthopteran jumping and flight: sensory basis of hindlimb movements. *Soc. Neurosci. Abstr. 4: 361* **flight windh SO mech**

ARBAS EA 1980 A neuroethological study of flight loss and aerial manoeuvering in insects. *Dissertation, University of Oregon, USA* **flight behav**

ARBAS EA 1983 Neuronal correlates of flight loss in a mexican grasshopper, *Barytettix psolus*: I. Motor and sensory cells. *J. comp. Neurol. 216: 369-380* **flight MN SR**

ARBAS EA 1983 Neuronal correlates of flight loss in a mexican grasshopper, *Barytettix psolus*: II. DCMD and TCG interneurons. *J. comp. Neurol. 216: 381-389* **flight DN**

ARBAS EA 1983 Aerial manoeuvering reflexes in flightless grasshoppers. *J. exp. Biol. 107: 509-514* **flight DN behav**

ARBAS EA 1983 Thoracic morphology of a flightless mexican grasshopper, *Barytettix psolus*: comparison with the locust, *Schistocerca gregaria. J. Morphol. 176: 141-153* **anat thor flight musc**

ARBAS EA 1986 Control of hindlimb posture by wind-sensitive hairs and antennae during locust flight. *J. comp. Physiol. A 159: 849-857* **flight windh mech ant SO leg**

ARBAS EA 1988 On the "flight" system of a flightless grasshopper. *Proc. 18th. Int. Congr. Entomol., Vancouver, Canada: 71* **flight**

ARBAS EA 1988 Transneuronal induction of muscle atrophy in adult grasshoppers. *Proc. 18th. Int. Congr. Entomol., Vancouver, Canada: 76* **leg musc autot degen**

ARBAS EA 1988 Atrophy of thoracic muscles follows hindlimb autotomy in grasshoppers. *Soc. Neurosci. Abstr. 14: 687* **leg musc autot**

ARBAS EA 1989 On the "flight" system of flightless grasshoppers. *In: Erber J Menzel Pflüger H-J Todt D (eds.), Neural mechanisms of behavior. Proc. 2nd Int. Congr. Neuroethology. Stuttgart & New York: Thieme Verlag. 122-123* **flight**

ARBAS EA 1989 Elements of the ancestral flight mechanism in secondarily flightless grasshoppers: homologues of the "wing-hinge stretch receptors". *Soc. Neurosci. Abstr. 15: 1287* **flight SR anat**

ARBAS EA TOLBERT LP 1986 Presynaptic terminals persist following degeneration of "flight" muscle during development of a flightless grasshopper. *J. Neurobiol. 17: 627-636* **degen NM flight**

ARBAS EA WEIDNER MH 1991 Trans-neuronal induction of muscle atrophy in grasshoppers. *J. Neurobiol. 22: 536-546* **autot leg degen musc**

ASHFORD M KERRY CJ KITS KS RAMSEY RL SANSOM MSP USHERWOOD PNR 1984 Kinetic analysis of channels gated by glutamate receptors in locust muscle membrane. *In: Cottrell GA (ed.), Electropharmacology of the in vitro synapse. St. Andrews: University Press. 14-15* **TS Glu musc channel recept**

ASHFORD M KERRY CJ KITS KS RAMSEY RL SANSOM MSP USHERWOOD PNR 1984 Kinetic analysis of channels gated by glutamate receptors in locust muscle membrane. *In: Tomczyk N (ed.) Biophysics of membrane transport. Wroclaw, Poland: University of Wroclaw Press. 157-166* **TS Glu musc channel recept**

ASHFORD MLJ BODEN P RAMSEY RL SANSOM MSP SHINOZAKI H USHERWOOD PNR 1988 Voltage-dependent block of locust muscle glutamate channels by chlorisondamine. *J. exp. Biol. 134: 131-154* **glu NM TS tox channel**

ASHFORD MLJ BODEN P RAMSEY RL SHINOZAKI H USHERWOOD PNR 1984 Voltage-dependent block by chlorisondamine of glutamate channels in *Schistocerca gregaria* muscle fibres. *J. Physiol. (Lond.) 357: 65P* **TS glu musc channel tox**

ASHFORD MLJ BODEN P RAMSEY RL SHINOZAKI H USHERWOOD PNR 1987 Effects of trimetaphan on locust muscle glutamate receptors. *J. exp. Biol. 130: 405-424* **TS glu musc tox recept**

ASHFORD MLJ BODEN P RAMSEY RL USHERWOOD PNR 1989 Enhancement of desensitization of quisqualate-type glutamate receptor by the dissociative anaesthetic ketamine. *J. exp. Biol. 141: 73-86* **TS glu desens recept tox**

ASHHURST DE 1965 The connective tissue sheath of the locust nervous system: its development in the embryo. *Quart. J. micr. Sci. 106: 61-73* **sheath devel**

ASHHURST DE 1987 Mucosubstances of the central nervous system: their distribution and histochemical characterization. *In: Gupta AP (ed.), Arthropod brain: its evolution, development, structure and function. New York: John Wiley & Sons Inc. 513-527* **sheath fineStruc**

ASHHURST DE CHAPMAN JA 1961 The connective tissue sheath of the nervous system of *Locusta migratoria*: an electron microscope study. *Quart. J. micr. Sci. 102: 463-7* **sheath fineStruc**

ATZINGER L 1957 Vergleichende Untersuchungen über die Beziehung zwischen Ausbildung der Flügel, der Flugmuskulatur und des Flugvermögens bei Feldheuschrecken. *Zool. Jb. (Anat.) 76: 199-222* **flight mus anat wing**

AUBELE E KLEMM N 1978 Origin, destination and mapping of tritocerebral neurones of locust. *Cell Tiss. Res. 178: 99-122* **anat suboes brain SGG mpt oc**

AUDSLEY N PHILIPS JE 1990 Stimulants of ileal salt transport in neuroendocrine system of desert locust. *Gen. comp. Endocrinol. 80: 127-137* **gut pep hormX**

AUGER D FESSARD A 1928 Observations sur l'excitabilité de l'organe tympanique du criquet. *C. R. Seanc. Soc. Biol., Paris 99: 400-401* **aud SO tymp**

AUTRUM H 1940 Über Lautäusserungen und Schallwahrnehmungen bei Arthropoden. II. Das Richtungshören von *Locusta* und Versuch einer Hörtheorie für Tympanalorgane vom Locustiden-Typen. *Z. vergl. Physiol. 28: 326-352* **aud SO tymp**

AUTRUM H 1941 Über Gehör- und Erschütterungssinn bei Locustiden. *Z. vergl. Physiol. 28: 580-637* **aud SO tymp subg vibr**

AUTRUM H 1964 Anatomy and physiology of sound receptors in invertebrates. *In: Busnel RG (ed.): Acoustic behaviour of animals. Amsterdam: Elsevier Publishing Co. 414-433* **tymp aud SO**

AUTRUM H 1975 L'audition et l'ouie des insectes. *In: Grassé PP (ed.): Traité de Zoologie, Vol. VIII, Fasc. III. Paris: Masson & Cie. 668-741* **tymp aud SO**

AUTRUM H SCHNEIDER W 1948 Vergleichende Untersuchungen über den Erschütterungssinn der Insekten. *Z. vergl. Physiol. 31: 77-88* **subg SO mech vibr**

AUTRUM H SCHWARTZKOPFF J SWOBODA H 1961 Der Einfluss der Schallrichtung auf die Tympanalpotentiale von *Locusta migratoria* L. *Biol. Zentralbl. 80: 385-402* **SO tymp aud**

AZIZ SA 1958 Probable hygroreceptors in the desert locust *Schistocerca gregaria* Forskål (Orthoptera: Acrididae). *Indian J. Entomol. 19: 164-170* **SO ant hygr**

BAADER A 1987 The role of motor neurons of the abdominal longitudinal muscles in flying locusts. *In: Elsner N Creutzfeld O (eds.), New frontiers in brain research: Proc. 15th Göttingen Neurobiol. Conf. Stuttgart, New York: Thieme Verlag. Abstract 52* **flight abd musc MN**

BAADER A 1988 Activity of neck motor neurons and flow-field modulated interneurons in flying locusts. *In: Elsner N Barth FG (eds.), Sense organs: interfaces between environment and behaviour: Proc.16th. Göttingen Neurobiol. Conf. Stuttgart, New York: Thieme Verlag. Abstract 121* **flight vis head abd MN wind DN**

BAADER A 1988 Some motor neurones of the abdominal longitudinal muscles of grasshoppers and their role in steering behaviour. *J. exp. Biol. 134: 455-462* **flight abd MN anat musc**

BAADER A 1989 Thoracic interneurons of *Locusta migratoria*: mediation of visually and acoustically induced orientation. *In: Elsner N Singer W (eds.), Dynamics and plasticity in neuronal systems: Proc. 17th. Göttingen Neurobiology Conference. Stuttgart, New York: Thieme Verlag. Abstract 146* **flight thor IN aud**

BAADER A 1990 The posture of the abdomen during locust flight: regulation by steering and ventilatory interneurones. *J. exp. Biol. 151: 109-131* **IN flight anat thor abd resp**

BAADER A 1990 An ascending visual neural pathway in locusts. *Naturwiss. 77: 338-340* **IN flight anat thor abd neck**

BAADER A 1991 Auditory interneurons in locusts produce directional head and abdomen movements. *J. comp. Physiol. A 169: 87-100* **flight DN IN thor behav aud**

BAADER A 1991 The contribution of some neck and abdominal motor neurons in locust steering reactions. *J. Insect Physiol. 37: 689-698* **flight MN neck abd aud rhythm**

BAADER A 1991 Simulation of self-motion in tethered flying insects: an optical flow field for locusts. *J. Neurosci. Methods 38: 193-199* **flight tech vis**

BACCETTI B 1955 Ricerche sulla fine struttura del perilemma nel sistema nervoso degli insetti. *Redia 40: 197-212* **sheath fineStruc**

BACON J 1980 A homologous interneurone in a locust, a cricket and a mantid. *Verh. dt. zool. Ges. 1980: 163* **var DN homol gryll dict**

BACON J 1983 Identified neurone approach to the study of insect flight. *In: Nachtigall W (ed.), BIONA-report 2 "Insekten Flug"; Akad. Wiss. Mainz. Stuttgart & New York: Gustav Fischer Verlag. 1-9* **flight DN**

BACON J ALTMAN JS 1977 A silver intensification method for cobalt-filled neurones in wholemount preparations. *Brain Res. 138: 359-363* **tech**

BACON J MÖHL B 1979 Activity of an identified wind interneurone in a flying locust. *Nature 278: 638-640* **DN flight**

BACON J MÖHL B 1983 The tritocerebral commissure giant (TCG) wind-sensitive interneuron in the locust. I. Its activity in straight flight. *J. comp. Physiol. A 150: 439-452* **DN flight**

BACON J TYRER NM 1978 The tritocerebral commissure giant (TCG): a bimodal interneurone in the locust *Schistocerca gregaria. J. comp. Physiol. A 126: 317-325* **DN anat**

BACON J TYRER NM 1979 Wind interneurone input to flight motor neurones in the locust *Schistocerca gregaria. Naturwiss. 66: 116* **DN MN flight**

BACON J TYRER NM 1979 The innervation of the wind-sensitive head hairs of the locust, *Schistocerca gregaria. Physiol. Entomol. 4: 301-309* **anat SO windh mech**

BACON JP TYRER NM HORSMANN U 1991 A putative inhibitory descending interneurone in the locust. *In: Bush BMM Armstrong BN (eds.): Neural and Sensory Mechanisms in Locomotion. Manchester: Manchester University Press. 37-48* **DN inhib GABA brain**

BADEN V 1936 Embryology of the nervous system in the grasshopper, *Melanoplus differentialis* (Acrididae: Orthoptera). *J. Morphol. 60: 159-188* **devel anat**

BADEN V 1938 Origin and fate of the median cord in the grasshopper, *Melanoplus differentialis* (Acrididae: Orthoptera). *J. Morphol. 63: 219-227* **devel**

BAEHR JC CARRUELLE JP PORAS M 1986 The activity of denervated corpora cardiaca in a diapausing strain of *Locusta migratoria*: in vivo and in vitro studies. *Int. J. Invert. Reprod. Devel. 10: 143-147* **retroc CC diap**

BAHADUR J ALI Z 1976 Nerve tracts in the metathoracic ganglion of the grasshopper *Acridella robusta* Uvarov (Orthoptera: Acrididae). *Zool. Jb. (Anat.) 96: 513-528* **anat thor**

BAILEY EV HARRIS MO 1991 Visual behaviour of the migratory grasshopper, *Melanoplus sanguinipes* F. (Orthoptera, Acrididae). *J. Insect Behav. 6: 707-726* **vis behav**

BAINES RA DOWNER RGH 1990 Serotonergic modulation of the locust mandibular closer muscle. *In: Borkovec AB Masler EP (eds.), Insect neurochemistry and neurophysiology. Clifton, N.J.: Humana Press Inc. 417-420* **musc mpt 5HT mod**

BAINES RA DOWNER RGH 1991 The role of proctolin in maintaining contractions of the locust (*Locusta migratoria*) mandibular closer muscle. *J. Insect Physiol. 37: 431-439* **mpt pep proct musc mod Ca2**

BAINES RA LANGE AB DOWNER RGH 1990 Proctolin in the innervation of the locust mandibular closer muscle modulates contraction through the elevation of inositol trisphosphate. *J. comp. Neurol. 297: 479-486* **pep proct innerv mpt musc suboes insP mod**

BAINES RA TYRER NM 1989 The innervation of locust salivary glands. II. Physiology of excitation and modulation. *J. comp. Physiol. A 165: 407-413* **saliv anat TS pep 5HT dopa norad proct FMRF MN suboes thor**

BAINES RA TYRER NM DOWNER RGH 1990 Serotonergic innervation of the locust mandibular closer muscle modulates contractions through the elevation of cyclic adenosine monophosphate. *J. comp. Neurol. 294: 623-632* **5HT innerv mpt musc cAMP suboes mod**

BAINES RA TYRER NM MASON JC 1989 The innervation of locust salivary glands. I. Innervation and analysis of transmitters. *J. comp. Physiol. A 165: 395-405* **saliv anat TS pep 5HT dopa norad proct FMRF MN suboes thor**

BAKER PS 1975 Optomotor responses of flying locusts. *Exp. Brain Res. 23: 13* **flight OM**

BAKER PS 1979 Flying locust visual responses in a radial wind tunnel. *J. comp. Physiol. A 131: 39-47* **flight OM tech**

BAKER PS 1979 The wing movements of flying locusts during steering behaviour. *J. comp. Physiol. A 131: 49-58* **flight aero**

BAKER PS 1979 The role of forewing muscles in the control of direction in flying locusts. *J. comp. Physiol. A 131: 59-66* **flight musc**

BAKER PS COOTER RJ 1979 The natural flight of the migratory locust *Locusta migratoria* L. I. Wing movements. *J. comp. Physiol. A 131: 79-88* **flight aero**

BAKER PS COOTER RJ 1979 The natural flight of the migratory locust *Locusta migratoria* L. II. Gliding. *J. comp. Physiol. A 131: 89-94* **flight aero**

BAKER PS GEWECKE M COOTER RJ 1981 The natural flight of the migratory locust *Locusta migratoria* L. III. Wing-beat frequency, flight speed and attitude. *J. comp. Physiol. A 141: 233-237* **flight aero**

BAKER PS GEWECKE M COOTER RJ 1984 Flight orientation of swarming *Locusta migratoria. Physiol. Entomol. 9: 247-252* **flight behav**

BALL EE BASTIANI MJ DE COUET HG QUINN JMA 1988 Transient patterned expression of the 1F8 antigen during development of the grasshopper nervous system. *Soc. Neurosci. Abstr. 14: 769* **devel**

BALL EE GOODMAN CS 1985 Muscle development in the grasshopper embryo: I. Muscles, nerves and apodemes in the metathoracic leg. *Dev. Biol. 111: 383-398* **musc devel leg**

BALL EE GOODMAN CS 1985 Muscle development in the grasshopper embryo: II. Syncytial origin of the extensor tibiae muscle pioneers. *Dev. Biol. 111: 399-416* **musc devel leg**

BALL EE GOODMAN CS 1985 Muscle development in the grasshopper embryo: III. Sequential origin of the flexor tibiae muscle pioneers. *Dev. Biol. 111: 417-424* **musc devel leg**

BALL EE HO RK GOODMAN CS 1985 Development of neuromuscular specificity in the grasshopper embryo: guidance of motoneuron growth cones by muscle pioneers. *J. Neurosci. 5: 1808-1819* **devel NM MN musc**

BALL EE QUINN JMA 1988 Neuromuscular development in the orthopteroid insects. *Proc. 18th. Int. Congr. Entomol., Vancouver, Canada: 76* **devel musc MN NM**

BALL FG KERRY CJ RAMSEY RL SANSOM MSP USHERWOOD PNR 1988 The use of dwell time cross-correlation functions to study single-ion channel gating kinetics. *Biophys. J. 54: 309-320* **TS glu musc channel**

BANNER SE CATTELL KJ OSBORNE RH 1986 Characterisation of 5-HT receptors in the foregut of the locust *Schistocerca gregaria. Brit. J. Pharmacol. 89: 565* **gut recept 5HT**

BANNER SE CATTELL KJ OSBORNE RH 1986 Effects of antagonists on 5-HT and octopamine receptors in the foregut of *Schistocerca gregaria. Brit. J. Pharmacol. 89: 825 P* **gut recept 5HT oct**

BANNER SE OSBORNE RH 1989 FMRFamide potentiation of 5-HT induced relaxation of the locust foregut is naloxone sensitive. *Brit. J. Pharmacol. 97: 516 P* **pep proct gut 5HT**

BANNER SE OSBORNE RH 1989 Modulation of 5-HT and proctolin receptors by FMRF-amide in the foregut of the locust *Schistocerca gregaria. J. Insect Physiol. 35: 887-892* **pep proct SGG gut mod 5HT FMRF recept**

BANNER SE OSBORNE RH CATTELL KJ 1987 The pharmacology of the isolated foregut of the locust *Schistocerca gregaria*. I. The effect of a range of putative neurotransmitters. *Comp. Biochem. Physiol. C 88: 131-138* **pep proct SGG gut 5HT oct ACh**

BANNER SE OSBORNE RH CATTELL KJ 1987 The pharmacology of the isolated foregut of the locust *Schistocerca gregaria*. II. Characterization of a 5HT2-like receptor. *Comp. Biochem. Physiol. C 88: 139-144* **SGG gut 5HT recept**

BANNER SE SEALE L OSBORNE RH CATTELL KJ 1988 Effects of proctolin on the fore and hindgut of the locust *Schistocerca gregaria*. *Pest. Sci. 23: 355-357* **pep proct gut**

BANNER SE WOOD SJ OSBORNE RH CATTELL KJ 1990 Tyramine antagonizes proctolin-induced contraction of the isolated foregut of the locust *Schistocerca gregaria* by an interaction with octopamine 2 receptors. *Comp. Biochem. Physiol. C 95: 233-236* **pep proct oct gut rhythm recept**

BAR-ILAN AR SHULOV A PENER MP 1969 The stridulation of sexually mature males of *Locusta migratoria migratorioides* (R. & F.) in relation to temperature and some other factors. *Physiol. Zool. 42: 417-428* **strid temp**

BARKER PDR 1987 Statistics of spike discharges from a visual unit in the locust. *Dissertation, Cambridge University, UK.* **DN vis sens TS**

BARONETZKY E MÖHL B 1987 Afferent input from the cerci on the locust flight motor. *In: Elsner N Creutzfeld O (eds.), New frontiers in brain research: Proc. 15th Göttingen Neurobiol. Conf. Stuttgart, New York: Thieme Verlag. Abstract 50* **flight cerc abd**

BARTON BROWNE L MOORHOUSE JE VAN GERWEN ACM 1975 Sensory adaptation and the regulation of meal size in the australian plague locust, *Chortoiocetes terminifera*. *J. Insect Physiol. 21: 1633-1639* **feed SO mpt**

BÄSSLER D WOLF H RATHMAYER W 1991 Long-term study of photoinactivation of an inhibitory motoneurone in the locust. *In: Elsner N Penzlin H (eds.), Synapse-transmission-modulation: Proc. 19th Göttingen Neurobiol. Conf. Stuttgart, New York: Thieme Verlag. Abstract 53* **inhib MN thor tech**

BÄSSLER U 1967 Zur Steuerung des Springens bei der Wanderheuschrecke *Schistocerca gregaria*. *Kybernetik 4: 112* **jump**

BÄSSLER U 1979 Effects of crossing the receptor apodeme of the femoral chordotonal organ on walking, jumping and singing in locusts and grasshoppers. *J. comp. Physiol. A 134: 173-176* **jump walk strid FCO SO propr**

BÄSSLER U 1983 Influence of femoral chordotonal organ afferences on ecdysis and on the development of motor programmes in locust larvae. *Physiol. Entomol. 8: 353-357* **moult FCO SO propr devel**

BÄSSLER U 1986 On the definition of central pattern generator and its sensory control. *Biol. Cybern. 54: 65-69* **flight walk rhythm**

BASTIANI MJ CARPENTER EM 1990 A novel subset of axon pathways in grasshopper embryos is labeled by the 10E6 mab. *Soc. Neurosci. Abstr. 16: 811* **devel**

BASTIANI MJ DOE CQ HELFLAND SL GOODMAN CS 1985 Neuronal specificity and growth cone guidance in grasshopper and *Drosophila* embryos. *Trends in Neurosci. 8: 257-266* **devel**

BASTIANI MJ DU LAC S GOODMAN CS 1985 The first growth cones in insect embryos: model system for studying the development of neuronal specificity. *In: Selverston AI (ed.), Model neural networks and behaviour. New York: Plenum Press. 149-174* **aud devel**

BASTIANI MJ DU LAC S GOODMAN CS 1986 Guidance of neuronal growth cones in the grasshopper embryo. I. Recognition of a specific axonal pathway by the pCC neuron. *J. Neurosci. 6: 3518-3531* **devel**

BASTIANI MJ GOODMAN CS 1984 The first growth cones in the central nervous system of the grasshopper embryo. *In: Black I (ed.), Cellular and molecular approaches to neuronal development. New York: Plenum Press. 63-84* **devel**

BASTIANI MJ GOODMAN CS 1984 Neuronal growth cones: specific interactions mediated by filopodial insertion and induction of coated vesicles. *Proc. nat. Acad. Sci. USA 81: 1849-1853* **devel**

BASTIANI MJ GOODMAN CS 1986 Guidance of neuronal growth cones in the grasshopper embryo. III. Recognition of specific glial pathways. *J. Neurosci. 6: 3541-3551* **devel glia**

BASTIANI MJ HARRELSON AL SNOW PM GOODMAN CS 1987 Expression of fasciclin I and II glycoproteins on subsets of axon pathways during neuronal development in the grasshopper. *Cell 48: 475-755* **devel**

BASTIANI MJ PEARSON KG GOODMAN CS 1984 From embryonic fascicles to adult tracts: organization of neuropile from a developmental perspective. *J. exp. Biol. 112: 45-64* **anat devel**

BASTIANI MJ RAPER JA GOODMAN CS 1984 Pathfinding by neuronal growth cones in grasshopper embryos. III. Selective affinity of the G growth cone for the P cells within the A/P fascicle. *J. Neurosci. 4: 2311-2328* **devel IN thor**

BATE CM 1976 Embryogenesis of an insect nervous system. I. A map of the thoracic and abdominal neuroblasts in *Locusta migratoria. J. Embryol. exp. Morphol. 35: 107-123* **devel IN MN thor abd homol**

BATE CM 1976 Pioneer neurones in an insect embryo. *Nature 260: 54-56* **devel SO**

BATE CM 1978 Development of sensory systems in arthropods. *In: Jacobson M (ed.), Handbook of sensory physiology, Vol IX. Berlin, Heidelberg, New York: Springer Verlag. 1-53* **devel SO**

BATE CM GOODMAN CS SPITZER NC 1981 Embryonic development of identified neurons: segment specific differences in the H cell homologues. *J. Neurosci. 1: 103-106* **aud devel DUM thor**

BATE CM GRUNEWALD EB 1981 Embryogenesis of an insect nervous system. II. A second class of precursor cells and the origin of the intersegmental connectives. *J. Embryol. exp. Morphol. 61: 317-330* **devel IN MN DUM**

BATEMAN A BODEN P DELL A DUCE IR QUICKE DLJ USHERWOOD PNR 1985 Postsynaptic block of a glutamatergic synapse by low molecular weight fractions of spider venom. *Brain Res. 339: 237-244* **TS glu NM channel tox**

BATES SE SANSOM MSP BALL FG RAMSEY RL USHERWOOD PNR 1990 Glutamate receptor-channel gating - maximum likelihood analysis of gigaohm seal recordings from locust muscle. *Biophys. J. 58: 219-229* **TS glu musc channel recept**

BAUER CK 1987 The deutocerebrum as motor center of the locust brain. *In: Elsner N Creutzfeld O (eds.), New frontiers in brain research: Proc. 15th Göttingen Neurobiol. Conf. Stuttgart, New York: Thieme Verlag. Abstract 77* **ant MN brain**

BAUER CK 1988 Neuronale Kontrolle der Antennenbewegung bei der Wanderheuschrecke *Locusta migratoria migratorioides* (Reiche & Fairmaire 1850). *Dissertation, Universität Hamburg, Germany.* **ant MN anat brain**

BAUER CK 1989 Neuropeptide proctolin modulates the excitation-contraction coupling in locust antennal muscles. *Progress Zool. 37: 371-372* **pep proct ant musc mod NM**

BAUER CK 1991 Modulatory action of proctolin in the locust (*Locusta migratoria*) antennal motor system. *J. Insect Physiol. 37: 663-674* **pep proct ant brain MN anat musc mod**

BAUER CK FISCHER-LOUGHEED J O'SHEA M 1988 Mapping of proctolin-like immunoreactivity in the locust brain and evidence for the presence of proctolin in antennal motoneurons. *In: Elsner N Barth FG (eds.), Sense organs: interfaces between environment and behaviour: Proc.16th. Göttingen Neurobiol. Conf. Stuttgart, New York: Thieme Verlag. Abstract 300* **pep proct ant brain MN anat**

BAUER CK GEWECKE M 1991 Motoneuronal control of antennal muscles in *Locusta migratoria. J. Insect Physiol. 37: 551-562* **ant brain MN anat musc innerv**

BAUER M VON HELVERSEN O 1987 Separate localization of sound recognizing and sound producing neural mechanisms in a grasshopper. *J. comp. Physiol. A 161: 95-101* **strid aud temp**

BAYRAMOGLU-ERGENE S 1964 Untersuchungen über den Einfluss der Ocellen auf die Fluggeschwindigkeit der Wanderheuschrecke *Schistocerca gregaria. Z. vergl. Physiol. 48: 467-480* **oc flight vis**

BAYRAMOGLU-ERGENE S 1968 Untersuchungen über die dynamogene Funktion der Antennen bei *Anacridium aegyptium. Z. vergl. Physiol. 52: 363-369* **ant flight walk**

BECK C GEWECKE M 1991 Control of the visual field by lobula neurons in locusts. *In: Elsner N Penzlin H (eds.), Synapse-transmission-modulation: Proc. 19th Göttingen Neurobiol. Conf. Stuttgart, New York: Thieme Verlag. Abstract 597 A* **vis IN anat opt**

BECKERT R 1972 Die larvalen Tympanalorgane der Wanderheuschrecke *Locusta migratoria migratorioides* (R. & F.) (Orthoptera: Acrididae). *Z. angew. Entomol. 70: 289-296* **aud SO tymp devel**

BEENAKKERS AMT BLOEMEN REB DE VLIEGER TA VAN DER HORST DJ VAN MARREWIJK WJA 1985 Insect adipokinetic hormones. *Peptides (Fayetteville) 6, suppl. 3: 437-444* **AKH pep rev**

BEENAKKERS AMT VAN DER HORST DJ VAN MARREWIJK JA 1985 Biochemical progresses directed to flight muscle metabolism. *In: Kerkut GA Gilbert LI (eds.): Comprehensive insect physiology, biochemistry and pharmacology. Vol. 10. London: Pergamon Press. 451-485* **musc flight metab**

BEENAKKERS AMT VAN DER HORST DJ VAN MARREWIJK WJA 1985 Insect lipids and lipoproteins, and their role in physiological processes. *Prog. Lipid. Res. 24: 19-67* **lip rev**

BEHRMANN P LAKES-HARLAN R HUSTERT R 1991 Development of the "Bauplan" of the mesothoracic peripheral nervous system in grasshoppers. *In: Elsner N Penzlin H (eds.), Synapse-transmission-modula-*

tion: Proc. 19th Göttingen Neurobiol. Conf. Stuttgart, New York: Thieme Verlag. Abstract 512 **devel thor**

BELANGER JH ORCHARD I 1987 Association of a proctolin-like peptide with the ovipositor muscle of *Locusta migratoria. Soc. Neurosci. Abstr. 13: 125* **pep proct abd genit musc TS**

BELANGER JH ORCHARD I 1988 Proctolinergic modulation of neuromuscular transmission at an ovipositor muscle of *Locusta migratoria. Soc. Neurosci. Abstr. 14: 534* **pep proct abd genit musc TS mod rhythm**

BELANGER JH ORCHARD I 1991 The problem of reliability in insect muscular control. *Soc. Neurosci. Abstr. 17: 1228* **abd genit musc**

BEN KHAY A MOREAU R GOURDOUX L PUIROUX J DUTRIEU J 1987 Les voies de dégradation du glucose chez le criquet migrateur: effets d'une neurohormone. *In: Physiologie, 55e. Réuníon Assoc. Physiol. Bordeaux 6-8 juillet 1987. Amsterdam: Elsevier Science Publishers BV. 38A* **metab hormX**

BENEDECZKY I 1978 Ultrastructural and cytochemical studies on the central body of *Locusta migratoria* L. *Proc. 2nd Europ. Neurosci. Meeting, Florence. 267* **brain CB fineStruc**

BENEDECZKY I 1981 The morphological features and physiological significance of the peptidergic neurosecretory neurons in insects. *Acta Agronom. Acad. Sci. Hung. 30: 305-322* **NSC**

BENEDECZKY I 1983 Ultrastructure of the neuromuscular junction in the hindgut of *Periplaneta americana* and *Locusta migratoria. Z. mikr.-anat. Forsch. (Leipzig) 97: 505-519.* **NM gut fineStruc dict**

BENEDECZKY I HALASY K CSOKNYA M 1987 Fine structure of the neuromuscular junctions in the alimentary tract of phylogenetically different animal species. *Acta biol. Hung. 38: 363-382* **gut NM fineStruc**

BENEDECZKY I KISS JZ SOMOGYI P 1982 Light and electron microscopic localization of substance P-like immunoreactivity in the cerebral ganglion of locusts with a monoclonal antibody. *Histochem. 75: 123-131* **pep SubP brain**

BENEDECZKY I S-RÓZSA K 1981 Cytochemical localization of adenylate cyclase in the various tissues of *Locusta migratoria (migratorioides* R. F.). *Histochem. 70: 179-188* **enz cAMP**

BENEDECZKY I 1981 The ultrastructure and cytochemistry of aminergic and peptidergic terminals in *Locusta migratoria migratorioides* R. F. *In: Rózsa KS (ed.): Adv. Physiol. Sci. Vol. 22: Neurotransmitters in invertebrates. Akadémiai Kiadó Budapest. Pergamon Press. 509-523* **fineStruc pep**

BENEDECZKY S-RÓZSA K 1981 Cytochemical demonstration of cyclic 3', 5' AMP phosphodiesterase in different tissue of migratory locust (*Locusta migratoria migratorioides* R. F.). *Histochem. 70: 189-197* **enz cAMP**

BENKE D BREER H 1988 Expression of neuronal acetylcholine receptor polypeptides. *Cell. mol. Neurobiol. 7: 391-401* **recept TS ACh**

BENKE D BREER H 1989 Endogenous low molecular weight inhibitors of cholinergic binding sites. *Insect Biochem. 19: 423-431* **TS ACh recept**

BENKE D BREER H 1990 Comparison of acetylcholine and alpha-bungarotoxin binding sites in insects and vertebrates. *Comp. Biochem. Physiol. C 94: 71-80* **TS ACh**

BENKE D STAHMER I BREER H 1989 Monoclonal antibodies detecting regulatory polypeptides of the insect neuronal acetylcholine receptor. *J. exp. Biol. 147: 329-341* **TS ACh recept**

BENKE D BREER H 1990 Identification of an endogenous polypeptide modulating ligand binding sites of insect neuronal acetylcholine receptors. *Neurochem. Int. 16 : 287-294* **pep ACh recept**

BENNETT RR 1966 Spectral sensitivities of locust retinula cells. *Am. Zool. 6: 520-521 (Abstract)* **SO vis ret**

BENNETT RR TUNSTALL J HORRIDGE GA 1967 Spectral sensitivities of single retinula cells of the locust. *Z. vergl. Physiol. 55: 195-206* **ret SO vis**

BENNET-CLARK HC 1975 The energetics of the jump of the locust *Schistocerca gregaria. J. exp. Biol. 63: 53-83* **jump**

BENNET-CLARK HC 1976 Energy storage in jumping animals. *In: Davies PS (ed.), Perspectives in experimental biology Vol I, Zoology. Oxford, New York::Pergamon. 467-479* **jump**

BENNET-CLARK HC ALDER BM 1979 The effect of air resistance on the jumping performance of insects. *J. exp. Biol. 82: 105-122* **jump**

BENNET-CLARK HC 1990 Jumping in Orthoptera. *In: Chapman RF Joern A (eds.), Biology of grasshoppers. New York: J. Wiley & Sons Inc. 173-204* **jump**

BENSON JA 1988 Bicuculline blocks the response to acetylcholine and nicotine but not to muscarine or GABA in isolated insect neuronal somata. *Brain Res. 458: 65-71* **ACh GABA TS thor tox**

BENSON JA 1988 Pharmacology of a locust thoracic ganglion somal nicotinic acetylcholine receptor. *In: Clementi F Gotti C Sher E (eds.), NATO ASI Series H, Vol. 25: Nicotinic acetylcholine receptors in the nervous system.*

Berlin, Heidelberg, New York: Springer Verlag. 227-240 **Ach TS thor recept**

BENSON JA 1988 Transmitter receptors on insect neuronal somata: GABAergic and cholinergic pharmacology. *In: Lunt GG (ed.): Neurotox '88: Molecular basis of drug and pesticide action. Amsterdam: Elsevier Science Publishers BV. 193-206* **ACh GABA TS thor recept**

BENSON JA 1989 Insect nicotinic acetylcholine receptors as targets for insecticides. *In: McFarlane NR (ed.), Progress and prospects in insect control. BCPC Monograph 43. Farnham: British Crop Protection Council. 59-70* **TS ACh recept**

BENSON JA 1989 M1-like muscarinic receptors mediate cholinergic activation of an inward current in isolated neuronal somata from locust thoracic ganglia. *Soc. Neurosci. Abstr. 15: 365* **ACh TS thor channel recept**

BENSON JA 1990 Nitromethylene heterocycle insecticides are antagonists at insect neuronal nicotinic receptors. *In: Frehse H Kesseler-Schmitz E Conway S (eds.), Book of Abstracts. Hamburg: Seventh Internat. Congr. Pesticide Chemistry. IUPAC 1: 348* **TS ACh recept tox**

BENSON JA NEUMANN R 1987 Nicotine and muscarine evoke different responses in isolated, neuronal somata from locust thoracic ganglia. *Soc. Neurosci. Abstracts 13: 938* **Ach TS thor**

BENTLEY DR 1969 A topological map of the locust flight system motor neurons. *J. Insect Physiol. 16: 905-918* **anat flight MN**

BENTLEY DR 1973 Postembryonic development of insect motor systems. *In: Young D (ed.), Developmental neurobiology of insect motor systems. Cambridge: Cambridge University Press. 147-177* **devel**

BENTLEY DR CAUDY M 1983 Navigational substrates for peripheral pioneer growth cones: limb-axis polarity cues, limb-segment boundaries, and guidepost neurons. *Cold Spring Harbor Symp. Quant. Biol. 48: 573-585* **devel**

BENTLEY DR KESHISHIAN H 1982 Pathfinding by peripheral pioneer neurons in grasshoppers. *Science 218: 1081-1088* **devel leg**

BENTLEY DR KESHISHIAN H SHANKLAND M TOROIAN-RAYMOND A 1979 Quantitative staging of embryonic development of the grasshopper *Schistocerca nitens. J. Embryol. exp. Morphol. 54: 47-74* **devel**

BENTLEY DR TOROIAN-RAYMOND A 1981 Embryonic and postembryonic morphogenesis of a grasshopper interneuron. *J. comp. Neurol. 201: 507-518* **devel DN brain**

BENTLEY DR TOROIAN-RAYMOND A 1989 Pre-axonogenesis migration of afferent pioneer cells in the grasshopper embryo. *J. exp. Zool. 251: 217-223* **devel**

BERÀNEK R MILLER PL 1968 The action of iontophoretically applied glutamate on insect muscle fibres. *J. exp. Biol. 49: 83-93* **musc leg NM TS glu**

BERLOT J GOODMAN CS 1984 Guidance of peripheral pioneer neurons in the grasshopper: an adhesive hierarchy of epithelial and neuronal surfaces. *Science 223: 293-295* **devel**

BERMUDEZ I BEADLE DJ BENSON JA 1990 Multiple serotonin-activated currents in isolated neuronal somata from locust thoracic ganglia. *Soc. Neurosci. Abstr. 16: 857* **TS 5HT cellCult**

BERNARD F 1937 Recherches sur la morphogenèse des yeux composées d'arthropodes. Développement, croissance, réduction. *Biol. Bull. Fr. Belg. Suppl. 23: 1-162* **opt anat devel**

BERNAYS EA 1970 Hatching in *Schistocerca gregaria* (Forskål): processes occurring between initial deposition and final hardening of the first instar cuticle. *Dissertation, University of London, UK.* **moult**

BERNAYS EA 1972 The muscles of newly hatched *Schistocerca gregaria* larvae and their possible function in hatching, digging and ecdysial movements (Insecta: Acrididae). *J. Zool. (Lond.) 166: 141-158* **musc anat moult**

BERNAYS EA 1972 The intermediate moult (first ecdysis) of *Schistocerca gregaria* (Forskål) (Insecta, Orthoptera). *Z. Morphol. Tiere 71: 160-179* **moult**

BERNAYS EA 1977 The physiological control of drinking behaviour in nymphs of *Locusta migratoria. Physiol. Entomol. 2: 261-273* **feed**

BERNAYS EA 1980 The post-branchial rest in *Locusta migratoria* and its hormonal regulation. *J. Insect Physiol. 26: 119-123* **feed behav hormX**

BERNAYS EA BLANEY WM CHAPMAN RF 1972 Changes in chemoreceptor sensilla on the maxillary palps of *Locusta migratoria* in relation to feeding. *J. exp. Biol. 57: 745-753* **feed SO chem mpt**

BERNAYS EA BLANEY WM CHAPMAN RF 1975 The problems of perception of leaf surface chemicals by locust chemoreceptors. *In: Denton DA Coghlan JP (eds.), Olfaction and taste V. New York: Academic Press. 227-230* **feed behav SO chem**

BERNAYS EA BLANEY WM CHAPMAN RF COOK AG 1976 The ability of *Locusta migratoria* to perceive plant surface waxes. *Symp. Biol. Hung. 16: 36-40* **feed SO chem**

BERNAYS EA CHAPMAN RF 1972 The control of changes in peripheral sensilla associated with feeding in *Locusta migratoria* (L.) *J. exp. Biol. 57: 755-764* **feed SO chem SGG brain retroc mpt**

BERNAYS EA CHAPMAN RF 1973 The regulation of feeding in *Locusta migratoria*: internal inhibitory mechanisms. *Entomol. exp. appl. 16: 329-342* **feed**

BERNAYS EA CHAPMAN RF 1974 The regulation of food intake by acridids. *In: Barton Browne L (ed.), Experimental analysis of insect behaviour. Berlin, Heidelberg, New York: Springer Verlag. 48-59* **feed**

BERNAYS EA CHAPMAN RF 1974 Changes in haemolymph osmotic pressure in *Locusta migratoria* larvae in relation to feeding. *J. Entomol. 48: 149-155* **feed blood**

BERNAYS EA MORDUE (LUNTZ) AJ 1973 Changes in the palp tip sensilla of *Locusta migratoria* in relation to feeding: the effects of different levels of hormone. *Comp. Biochem. Physiol. A 45: 451-454* **feed SO mpt hormX chem**

BERNAYS EA SIMPSON SJ 1982 Control of food intake. *Adv. Insect Physiol. 16: 59-118* **feed behav SO**

BERNAYS EA WRUBEL RP 1985 Learning by grasshoppers: association of colour/light intensity with food. *Physiol. Entomol. 10: 359-369* **feed behav learn vis**

BHARADWAJ RK BANERJEE SK 1970 Anatomy and histology of the circulatory system of the desert locust, *Schistocerca gregaria* (Orthoptera: Acrididae). *J. East Afr. Nat. Hist. Soc. Nat. Mus. 4 : 329-345* **anat heart**

BHARADWAJ RK BANERJEE SK 1971 The nervous system of the desert locust, *Schistocerca gregaria* (Orthoptera: Acrididae) with a discussion on muscle innervation. *J. nat. Hist. 5: 183-208* **anat musc flight innerv**

BICKELY WME 1942 On the stomodaeal nervous system of insects. *Ann. entomol. Soc. Am. 35: 343-354* **SGG anat**

BICKER G 1991 Taurine-like immunoreactivity in photoreceptor cells and mushroom bodies: a comparison of the chemical architecture of insect nervous systems. *Brain Res. 560: 201-206* **TS tau opt brain oc MB anat ret**

BICKER G PEARSON KG 1983 Initiation of flight by an identified wind sensitive neurone (TCG) in the locust. *J. exp. Biol. 104: 289-294* **DN flight**

BICKER G SCHÄFER S OTTERSEN OÜ STORM-MATHISEN J 1988 Glutamate-like immunoreactivity in identified neuronal populations of insect nervous systems. *J. Neurosci. 8: 2108-2122* **TS Glu brain**

BICKER G KIRCHHOF B KREISSL S 1989 Identified populations of insect neurons in the petri dish. *In: Erber J Menzel R Pflüger H-J Todt D (eds.), Neural mechanisms of behavior: Proc. 2nd Internat. Congr. Neuroethology. Stuttgart & New York: Thieme Verlag. Abstract 247* **cellCult**

BICKMEYER U AHI J 1990 Physiology and morphology of vibrosensitive neurones in the thoracic ganglia of the migratory locust. *In: Elsner N Roth G (eds.) Brain - Perception - Cognition: Proc. 18th Göttingen Neurobiol. Conf. Stuttgart, New York: Thieme Verlag. Abstract 529* **vibr IN thor anat**

BISHOP CA O'SHEA M DINERSTEIN RJ 1980 Techniques for detecting amines and tyrosine containing peptides in the CNS of locusts and other insects. *Soc. Neurosci. Abstr. 6: 627* **tech pep**

BLACKITH RE BLACKITH RM 1966 The anatomy and physiology of the morabine grasshoppers. II. External anatomy and comparisons with Pyrgomorphidae, Acrididae, and Proscopiidae. *Aust. J. Zool. 14: 1035-1075* **anat**

BLACKITH RE BLACKITH RM 1966 The anatomy and physiology of the morabine grasshoppers. I. Digestive and reproductive systems. *Aust. J. Zool. 14: 31-48* **gut genit anat mor**

BLACKITH RE BLACKITH RM 1967 The anatomy and physiology of the morabine grasshoppers. III. Muscles, nerves, tracheae and genitalia. *Aust. J. Zool. 15: 961-998* **anat musc genit mor**

BLACKITH RE GOTO HE 1974 Antennal structures of morabine grasshoppers. *Acrida 3: 7-17* **SO ant anat mor**

BLAND RG 1982 Morphology and distribution of sensilla on the antennae and mouthparts of Hypochlora alba (Orthoptera: Acrididae). *Ann. entomol. Soc. Am. 75: 272-283* **ant chem feed SO anat**

BLAND RG 1989 Antennal sensilla of Acrididae (Orthoptera) in relation to subfamily and food preference. *Ann. entomol. Soc. Amer. 82: 368-384* **ant chem feed SO**

BLANEY WM 1974 Electrophysiological responses of the terminal sensilla on the maxillary palps of *Locusta migratoria* (L.) to some electrolytes and non-electrolytes. *J. exp. Biol. 60: 275-293* **SO mpt chem**

BLANEY WM 1975 Behavioural and electrophysiological studies of taste discrimination by the maxillary palps of larvae of *Locusta migratoria. J. exp. Biol. 62: 555-570* **SO mpt chem behav**

BLANEY WM 1977 The ultrastructure of an olfactory sensillum on the maxillary palps of *Locusta migratoria. Cell & Tissue Res. 184: 397-409* **SO chem mpt fineStruc**

BLANEY WM 1980 Chemoreception and food selection by locusts. *In: Van Der Starre H (ed.) Olfaction and taste, Vol 7. London: Information Retrieval Limited. 127-130* **mpt feed SO chem behav**

BLANEY WM 1981 Chemoreception and food selection in locusts. *Trends in Neurosci. 4: 35-58* **mpt feed SO chem behav**

BLANEY WM CHAPMAN RF 1969 The anatomy and histology of the maxillary palp of *Schistocerca gregaria* Forskål (Orthoptera, Acrididae). *J. Zool. (Lond.) 157: 509-535* **SO mpt anat CO**

BLANEY WM CHAPMAN RF 1969 The fine structure of the terminal sensilla on the maxillary palps of *Schistocerca gregaria* (Forskål) (Orthoptera: Acrididae). *Z. Zellforsch. mikr. Anat. 99: 74-97* **SO mpt fineStruc moult**

BLANEY WM CHAPMAN RF 1970 The functions of the maxillary palps of Acrididae (Orthoptera). *Entomol. exp. appl. 13: 363-376* **mpt feed behav**

BLANEY WM CHAPMAN RF COOK AG 1971 The structure of the terminal sensilla on the maxillary palps of *Locusta migratoria* (L.) and changes associated with moulting. *Z. Zellforsch. mikr. Anat. 121: 48-68* **SO mpt anat fineStruc**

BLANEY WM CHAPMAN RF WILSON A 1973 The pattern of feeding of *Locusta migratoria* (L.) (Orthoptera, Acrididae). *Acrida 2: 119-137* **feed behav**

BLANEY WM DUCKETT A 1975 The significance of palpation by the maxillary palps of *Locusta migratoria* (L.): an electrophysiological and behavioural study. *J. exp. Biol. 63: 701-712* **SO mpt feed behav chem mech**

BLANEY WM SCHOONHOVEN LM SIMMONDS MSJ 1986 Sensitivity variations in insect chemoreceptors: a review. *Experientia 42: 13-19* **mpt SO devel rev**

BLANEY WM SIMMONDS MSJ 1983 Electrophysiological activity in insects in response to antifeedants. *Tropical Development and Research Institute, London. 219 pp* **mpt feed behav chem mech**

BLANEY WM SIMMONDS MSJ 1985 Food selection by locusts: the role of learning in rejection behaviour. *Entomol. exp. appl. 39: 273-278* **feed learn behav**

BLANEY WM SIMMONDS MSJ 1987 Control of mouthparts by the suboesophageal ganglion. *In: Gupta AP (ed.), Arthropod brain: its evolution, development, structure and function. New York: John Wiley & Sons Inc. 303-322* **mpt feed suboes rev**

BLANEY WM SIMMONDS MSJ 1990 The chemoreceptors. *In: Chapman RF Joern A (eds.), Biology of grasshoppers. New York: J. Wiley & Sons Inc. 1-37* **SO chem ant mpt leg rev**

BLANEY WM SIMMONDS MSJ SIMPSON SJ 1990 Dietary selection behaviour: comparisons between locusts and caterpillars.*In: Proc. 7th Int. Meeting on Insect/Plant Relations. Symp. Biol. Hung. 39: 47-52* **feed behav**

BLANEY WM SIMMONDS MSJ EVANS SV FELLOWS LE 1984 The role of the secondary plant compound 2,5-dihydroxymethyl 3,4-dihydroxypyrrolidine as a feeding inhibitor of insects. *Entomol. exp. appl. 36: 209-216* **mpt feed So chem**

BLANEY WM WINSTANLEY C 1980 Chemosensory mechanisms in locusts in relation to feeding: the role of some secondary plant compounds. *In: Sherwood M (ed.): Insect neurobiology and pesticide action (Neurotox 79). London: Society of Chemical Industry Press. 383-389* **mpt feed SO chem**

BLAU C CANDY DJ WEGENER G 1991 Regulaton of fructose 2,6-bisphosphate in perfused flight muscle of the locust, *Locusta migratoria*: the effect of octopamine. *Biochem. Soc. Trans. 19: 136* **metab musc flight oct**

BLOEMEN REB BEENAKKERS AMT DE VLIEGER TA 1986 Influences of the NCC axons on electrical activity in the glandular lobe of the corpus cardiacum of *Locusta migratoria. Comp. Biochem. Physiol. A 84: 319-324* **retroc CC brain**

BLOEMEN REB DE VLIEGER TA 1985 Spontaneous electrical activity recorded extracellularly from the glandular lobe of the corpus cardiacum of *Locusta migratoria. Comp. Biochem. Physiol. A 82: 627-633* **retroc CC**

BLOEMEN REB DE VLIEGER TA BEENAKKERS AMT 1987 The inhibitory effect of trehalose on electrical activity in the glandular lobe of in vitro corpus cardiacum of *Locusta migratoria. Comp. Biochem. Physiol. A 87: 193-195* **retroc CC**

BLOOM JW ATWOOD HL 1980 Effects of altered sensory experience on the responsiveness of the locust descending contralateral movement detector neuron. *J. comp. Physiol. A 135: 191-199* **DN vis opt habit**

BLOOM JW ATWOOD HL 1981 Reversible ultrastructural changes in the rhabdom of the locust eye are induced by long-term light deprivation. *J. comp. Physiol. A 144: 357-365* **ret fineStruc**

BLUST MH HOPKINS TL 1987 Olfactory responses of a specialist and a generalist grasshopper to volatiles of *Artemisia ludoviciana* Nutt. (Asteraceae). *J. chem. Ecol. 13: 1893-1902* **ant feed chem SO**

BODEN P BYCROFT BW CHHABRA SR CHIPLIN J CROWLEY PJ GROUT RJ KING TJ MCDONALD E RAFFERTY P USHERWOOD PNR 1986 The action of natural and synthetic isomers of quisqualic acid at a well-defined glutamatergic synapse. *Brain Res. 385: 205-211* **TS glu NM channel tox**

BOECKHOFF I 1990 Molecular reaction cascades in olfactory signal transduction. *In: Elsner N Roth G (eds.) Brain - Perception - Cognition: Proc. 18th Göttingen Neurobiol. Conf. Stuttgart, New York: Thieme Verlag. Abstract 309* **lip ant SO chem**

BOEKH J 1967 Reaktionsschwelle, Arbeitsbereich und Spezifizität eines Geruchsrezeptors auf der Heuschreckenantenne. *Z. vergl. Physiol. 55: 378-406* **ant SO chem**

BOEKH J ERNST K.-D. 1987 Zur Topographie der primären Projektionen von Sinnesorganen des Mundbereichs im Insektengehirn. *In: Elsner N Creutzfeld O (eds.), New frontiers in brain research: Proc. 15th Göttingen Neurobiol. Conf. Stuttgart, New York: Thieme Verlag. Abstract 75* **SO mpt anat suboes**

BOEKHOFF I STROTMANN J RAMING K TAREILUS E BREER H 1990 Odorant-sensitive phopholipase C in insect antennae. *Cellular Signalling 2: 49-56* **SO chem recept enz ant**

BÖHME C BRÄUNING P 1990 New features of the retrocerebral glandular complex of the locust. *In: Elsner N Roth G (eds.) Brain - Perception - Cognition: Proc. 18th Göttingen Neurobiol. Conf. Stuttgart, New York: Thieme Verlag. Abstract 314* **retroc anat IN**

BOULTON PS 1969 Degeneration and regeneration in the insect central nervous system. I. *Z. Zellforsch. mikr. Anat. 107: 98-118* **degen regen**

BOULTON PS ROWELL CHF 1969 Degeneration and regeneration in the insect central nervous system. II. *Z. Zellforsch. mikr. Anat. 107: 119-134* **degen regen**

BOUREME D FOURNIER B MATZ G GIRARDIE J 1989 Immunological and functional cross-reactivities between locust neuroparsins and proteins from cockroach corpora cardiaca. *J. Insect Physiol. 35: 265-271* **neuropars dict pep**

BOUREME D GIRARDIE A GIRARDIE J 1985 Biosynthèse des protéines de neurosécrétion du système pars intercerebralis-corpora cardiaca du criquet migrateur. *Ann. Endocrinol. 46: 14-15* **pep NSC retroc CC**

BOUREME D GIRARDIE J 1986 Présence de protéines à ponts disulfure dans les corpora cardiaca d'insectes. *C. R. Acad. Sci. Paris Ser. D 302: 655-660* **retroc CC**

BOUREME D GIRARDIE J 1989 Les neuroparsines du criquet: microhétérogénéité d'origine enzymatique (Orthoptera). *Ann. Soc. entomol. Fr. N.S. 24: 390-391* **neuropars pep**

BOUREME D TAMARELLE M GIRARDIE J 1987 Production and characterization of antibodies to neuroparsins A and B isolated from the corpora cardiaca of the locust. *Gen. comp. Endocrinol. 67: 169-177* **pep neuropars retroc CC**

BOYAN GS 1983 Postembryonic development in the auditory system of the locust. Anatomical and physiological characterisation of interneurones ascending to the brain. *J. comp. Physiol. A 151: 499-513* **aud IN anat devel**

BOYAN GS 1984 Neural mechanisms of information processing by identified interneurones in Orthoptera. *J. Insect Physiol. 30: 27-41* **aud IN anat**

BOYAN GS 1984 What is an "auditory" neurone? *Naturwiss. 71: 482-484* **aud flight IN thor**

BOYAN GS 1985 Responses to sound of identified neurones in the flight system of the locust. *In: Kalmring K Elsner N (eds.) Acoustic and vibrational communication in insects. Hamburg & New York: Paul Parey Verlag. 82-88* **IN flight aud**

BOYAN GS 1985 Auditory input to the flight system of the locust. *J. comp. Physiol. A 156: 79-91* **flight aud IN**

BOYAN GS 1985 Response decrement in an auditory neurone of the locust. *J. Insect Physiol. 31: 99-107* **IN aud thor habit**

BOYAN GS 1986 Modulation of auditory responsiveness in the locust. *J. comp. Physiol. A 158: 813-825* **aud IN synon mod**

BOYAN GS 1987 Primary afferent depolarization, presynaptic inhibition, and information processing in the cercal system of the locust. *In: Elsner N Creutzfeld O (eds.), New Frontiers in brain research. Proc. 15th Göttingen*

Neurobiol. Conf. Stuttgart, New York: Thieme Verlag. Abstract 8 **cerc abd inhib IN GF**

BOYAN GS 1988 Presynaptic inhibition of identified wind-sensitive afferents in the cercal system of the locust. *J. Neurosci. 8: 2748-2757* **cerc abd mech windh SO inhib propr IN GF**

BOYAN GS 1988 The cellular basis of a behaviour: from sensory neurone to behavioural output in the flight system of the locust. *Neurosci. Lett. Suppl. 30: 24-25* **cerc abd flight IN GF rhythm**

BOYAN GS 1989 Is there a common "Bauplan" for insect auditory pathways? *In: Erber J Menzel Pflüger H-J Todt D (eds.), Neural mechanisms of behavior: Proc. 2nd Internat. Congr. Neuroethology. Stuttgart & New York: Thieme Verlag. Abstract 73a* **aud IN SO tymp thor lep**

BOYAN GS 1989 Synaptic input converging on to an identified premotor interneuron in the locust is gated according to modality. *Naturwiss. 76: 431-433* **IN DN thor flight**

BOYAN GS 1990 Integration of auditory information for avoidance flight in the grasshopper and noctuid moth. *In: Elsner N Roth G (eds.) Brain - Perception - Cognition: Proc. 18th Göttingen Neurobiol. Conf. Stuttgart, New York: Thieme Verlag. Abstract 72* **aud batEv IN lep**

BOYAN GS 1991 Synaptic properties of segmentally homologous interneurones in the locust. *In: Elsner N Penzlin H (eds.), Synapse-transmission-modulation: Proc. 19th Göttingen Neurobiol. Conf. Stuttgart, New York: Thieme Verlag. Abstract 515* **IN thor aud devel**

BOYAN GS 1991 A neuronal network defined by lineage in the locust CNS. *Naturwissenschaften 78: 565-567* **thor IN aud homol**

BOYAN GS ALTMAN JS 1985 The suboesophageal ganglion: a "missing link" in the auditory pathway of the locust. *J. comp. Physiol. A 156: 413-428* **aud vis IN suboes anat**

BOYAN GS BALL EE 1989 The wind-sensitive cercal receptor/giant interneurone system of the locust *Locusta migratoria*. II. Physiology of giant interneurons. *J. comp. Physiol. A 165: 511-521* **cerc GF IN abd anat**

BOYAN GS BALL EE 1989 Parallel inputs shape the response of a giant interneurone in the cercal system of the locust. *J. Insect Physiol. 35: 305-312* **cerc GF IN abd anat**

BOYAN GS BALL EE 1990 Neuronal organization and information processing in the wind-sensitive cercal receptor/giant interneurone system of the locust and other orthopteroid insects. *Prog. Neurobiol. 35: 217-243* **cerc abd windh SO mech GF IN**

BOYAN GS WILLIAMS JLD BALL EE 1989 The wind-sensitive cercal receptor/giant interneurone system of the locust *Locusta migratoria.* I. Anatomy of the system. *J. comp. Physiol. A 165: 495-510* **cerc GF IN abd anat synon**

BOYAN GS WILLIAMS JLD BALL EE 1989 The wind-sensitive cercal receptor/giant interneurone system of the locust *Locusta migratoria.* III. Cercal activation of thoracic motor pathways. *J. comp. Physiol. A 165: 523-537* **cerc GF IN abd anat flight**

BOYAN GS WILLIAMS JLD BALL EE 1989 The wind-sensitive cercal receptor/giant interneurone system of the locust *Locusta migratoria.* IV. The non-giant interneurones. *J. comp. Physiol. A 165: 539-552* **cerc IN abd anat**

BOYAN GS ASHMAN S BALL EE 1986 Initiation and modulation of flight by a single giant interneuron in the cercal system of the locust. *Naturwiss. 73: 272-274* **IN GF flight**

BOYD K EWER DW 1949 Flight responses of grasshoppers. *S. Afr. Sci. 2: 168-169* **windh SO mech flight**

BRÄUNIG P 1982 Distribution and specific central projections of mechanoreceptors in the thorax and proximal leg joints of locusts. III. The external mechanoreceptors: the campaniform sensilla. *Cell Tiss. Res. 222: 647-654* **SO mech thor leg**

BRÄUNIG P 1982 The peripheral and central nervous organization of the locust coxo-trochanteral joint. *J. Neurobiol. 13: 413-433* **anat mech SO leg thor**

BRÄUNIG P 1985 Mechanoreceptive neurons in an insect brain. *J. comp. Neurol. 236: 234-240* **brain SO propr ant mech**

BRÄUNIG P 1985 Strand receptors associated with the femoral chordotonal organs of locust legs. *J. exp. Biol. 116: 331-341* **FCO propr SO anat**

BRÄUNIG P 1987 Serotonin-immunoreactive neurosecretory cells of the locust suboesophageal ganglion. *In: Elsner N Creutzfeld O (eds.), New frontiers in brain research: Proc. 15th Göttingen Neurobiol. Conf. Stuttgart, New York: Thieme Verlag. 254* **NSC suboes 5HT TS**

BRÄUNIG P 1987 The satellite nervous system - an extensive neurohaemal network in the locust head. *J. comp. Physiol. A 160: 69-77* **mpt NSC suboes**

BRÄUNIG P 1988 Morphology of neuronal elements in the nervus corporis cardiaca III (NCC III) of locusts. *In: Elsner N Barth FG (eds.), Sense organs: interfaces between environment and behaviour: Proc.16th. Göttingen*

Neurobiol. Conf. Stuttgart, New York: Thieme Verlag. Abstract 293 **DUM suboes retroc ant oct TS**

BRÄUNIG P 1988 Neurosecretory cells of the locust suboesophageal ganglion. *In: Salánki J S-Rózsa K (eds.), Neurobiology of invertebrates (Symp. Biol. Hung. 36). Budapest: . 161-172* **suboes TS 5HT anat NSC**

BRÄUNIG P 1988 Identification of a single prothoracic "dorsal unpaired median" (DUM) neuron supplying locust mouthpart nerves. *J. comp. Physiol. A 163: 835-840* **DUM thor mpt suboes anat**

BRÄUNIG P 1990 The morphology of suboesophageal ganglion cells innervating the nervus corporis cardiaci III of the locust. *Cell Tiss. Res. 260: 93-108* **anat suboes retroc CAll DUM**

BRÄUNIG P 1990 The mandibular ganglion - a new peripheral ganglion of the locust. *J. exp. Biol. 148: 313-324* **anat suboes mpt SGG SO propr**

BRÄUNIG P 1991 The heart neurone of locusts. An example for the identification of an individual neurone by a combination of neuroanatomical, neurphysiological, and immunohistochemical methods. *In: Elsner N Penzlin H (eds.), Synapse-transmission-modulation: Proc. 19th Göttingen Neurobiol. Conf. Stuttgart, New York: Thieme Verlag. Abstract 613* **IN suboes pep heart**

BRÄUNIG P 1991 A suboesophageal ganglion cell innervates heart and retrocerebral glandular complex in the locust. *J. exp. Biol. 156: 567-582* **anat suboes retroc CC NSC pep BPP FMRF heart thor abd medN**

BRÄUNIG P 1991 Suboesophageal DUM neurones innervate the principal neuropiles of the locust brain. *Phil. Trans. R. Soc. Lond. (Biol.) 332: 221-240* **suboes DUM brain anat**

BRÄUNIG P ALLGÄUER C HONEGGER H-W 1990 Suboesophageal DUM neurones are part of the antennal motor system of locusts and crickets. *Experientia 46: 259-261* **DUM suboes ant MN retroc gut anat gryll**

BRÄUNIG P CAHILL MA HUSTERT R 1986 The coxo-trochanteral muscle receptor organ of locusts: dendritic tubular bodies in a non-ciliated insect mechanoreceptive neuron. *Cell Tiss. Res. 243: 517-524* **SO mech fineStruc leg**

BRÄUNIG P HUSTERT R 1980 Proprioceptors with central cell bodies in insects. *Nature 283: 768-770* **propr SO**

BRÄUNIG P HUSTERT R 1983 Proprioceptive control of a muscle receptor organ in the locust leg. *Brain Res. 274: 341-343* **SO propr leg MN**

BRÄUNIG P HUSTERT R 1985 Actions and interactions of proprioceptors of the locust hind leg coxo-trochanteral joint. I. Afferent responses in relation to joint position and movement. *J. comp. Physiol. A 157: 73-82* **SO leg propr MN**

BRÄUNIG P HUSTERT R 1985 Actions and interactions of proprioceptors of the locust hind leg coxo-trochanteral joint. II. Influence on the motor system. *J. comp. Physiol. A 157: 83-89* **SO leg propr MN**

BRÄUNIG P HUSTERT R PFLÜGER H-J 1981 Distribution and specific central projections of mechanoreceptors in the thorax and proximal leg joints of locusts. 1. Morphology, location and innervation of internal proprioceptors of pro- and metathorax and their central projections. *Cell Tiss. Res. 216: 57-77* **SO thor leg propr anat**

BRÄUNIG P PFLÜGER H-J HUSTERT R 1983 The specificity of central nervous projection of locust mechanoreceptors. *J. comp. Neurol. 218: 197-207* **SO mech anat mpt ant brain suboes thor**

BRÄUNIG P SCHACHTNER J 1991 In vivo recording from locust neurosecretory cells. *In: Elsner N Penzlin H (eds.), Synapse-transmission-modulation: Proc. 19th Göttingen Neurobiol. Conf. Stuttgart, New York: Thieme Verlag. Abstract 389* **NSC suboes 5HT**

BRECKOW J KALMRING K ECKHORN E 1982 Multi-channel recordings and real-time current source density (CSD) analysis in the central nervous system of insects. *Biol. Cybern. 45: 115-121* **aud IN SO**

BRECKOW J KEUPER A 1984 Conductivities and anisotropies in diffusely arranged nervous tissues; their determination and importance in neurophysiological studies. *Biophys. Struct. Mech. 10: 185-196* **tech**

BRECKOW J SIPPEL M 1985 Mechanics of the transduction of sound in the tympanal organ of adults and larvae of locusts. *J. comp. Physiol. A 157: 619-629* **SO tymp aud**

BREER H 1981 Comparative studies on cholinergic activities in the central nervous system of *Locusta migratoria. J. comp. Physiol. A 141: 271-275* **TS ACh**

BREER H 1981 Characterization of synaptosomes from the central nervous system of insects. *Neurochem. Int. 3: 155-163* **synaptosome**

BREER H 1982 Uptake of (N-Me-3H)-choline by synaptosomes from the central nervous system of *Locusta migratoria. J. Neurobiol. 13: 107-117* **synaptosome TS ACh**

BREER H 1983 Venoms and toxins in neurochemical research of insects. *In: Hucho F Ovchinnikov YA (eds.): Toxins as tools in neurochemical research on insects. Berlin: de Gruyter. 115-125* **tox**

BREER H 1986 Chemistry of synapses and synaptic transmission in the nervous system of insects. *In: Borkovec AB Gelman DB (eds.), Insect neurochemistry and neurophysiology. Clifton, N.J.: Humana Press. 91-115* **synaptosome TS ACh**

BREER H 1986 Synaptosomes - systems for studying insect neurochemistry. *In: Ford MG Lung GG Reay RC Usherwood PNR (eds.), Neuropharmacology and pesticide action. Chichester: Ellis Horwood. 384-413* **synaptosome**

BREER H 1987 Neurochemical aspects of cholinergic synapses in the insect brain. *In: Gupta AP (ed.), Arthropod brain: its evolution, development, structure and function. New York: John Wiley & Sons Inc. 415-438* **TS ACh**

BREER H 1987 Identification and reconstitution of neuronal acetylcholine receptors from insects. *In: Maelicke A (ed.): Mechanism of action in the nicotinic acetylcholine receptor. Berlin, Heidelberg, New York: Springer Verlag. 334-348* **recept TS ACh**

BREER H BENKE D 1985 Synthesis of acetylcholine receptors in *Xenopus* oocytes induced by poly(A) mRNA from locust nervous system. *Naturwiss. 72: 213-214* **TS ACh recept**

BREER H BENKE D 1986 Messenger RNA from insect nervous tissue induces expression of neuronal acetylcholine receptors in *Xenopus* oocytes. *Mol. Brain Res. 1: 111-117* **TS ACh recept**

BREER H BENKE D HANKE W WIECZOREK L 1987 Studies on insect neuronal transmitter receptors and their encoding nucleic acids. *In: Tucek J (ed.): Metabolism of the brain. New York: John Wiley & Sons. 167-173* **TS ACh recept**

BREER H BENKE D KNIPPER M KLEENE R HANKE W WIECZOREK L 1987 Isolation, reconstitution and expression of acetylcholine receptor proteins from the central nervous system of insects. *In: Law JH (ed.): Molecular entomology. New York: Alan R. Liss Inc. 95-106* **recept TS ACh**

BREER H BOEKHOFF I STROTMANN JR MAING K TAREILUS E 1990 Molecular elements of olfactory signal transduction in insect antennae. *In: Schild D (ed.): Chemosensory information processing. Berlin, Heidelberg, New York: Springer Verlag. 77-86* **SO chem recept**

BREER H HEILIGENBERG H 1985 Neurochemistry of GABAergic activities in the central nervous system of *Locusta migratoria*. *J. comp. Physiol. A 157: 343-354* **TS GABA**

BREER H JESERICH G 1981 Calcium binding sites of synaptosomes from insect nervous system as probed by trivalent terbium ions. *J. Neurochem. 37: 276-282* **Ca2 synaptosome**

BREER H JESERICH G 1984 Invertebrate synaptosomes - implications for comparative neurochemistry. *In: Jones DG (ed.) Current topics in research of synapses 1: 165-210* **TS ACh synaptosome**

BREER H KLEENE R HINZ G 1985 Molecular forms and subunit structure of the acetylcholine receptor in the central nervous system. *J. Neurosci. 5: 3386-3392* **TS ACh recept**

BREER H KNIPPER M 1983 Characterization of acetylcholine release from insect synaptosomes. *Insect Biochem. 14: 337-344* **TS ACh synaptosome**

BREER H KNIPPER M 1985 Effects of neurotoxins on the high affinity translocation of choline. *Comp. Biochem. Physiol. C 81: 219-222* **TS ACh tox**

BREER H KNIPPER M 1985 Synaptosomes and neuronal membranes from insects. *In: Breer H Miller TA (eds.): Neurochemical techniques in insect research. Berlin, Heidelberg, New York: Springer Verlag. 125-156* **TS ACh synaptosome**

BREER H KNIPPER M BENKE D HANKE W 1987 Properties of the high affinity choline carrier and acetylcholine receptors in the central nervous system of insects. *In: Hucho F Ovchinnikov YA (eds.): Receptors and ion channels. Berlin: de Gruyter. 89-97* **TS ACh recept**

BREER H LÖEEME R BENKE D 1984 Isolation of a putative nicotinic acetylcholine receptor from the central nervous system of *Locusta migratoria. Neurosci. Lett. 46: 323-328* **TS ACh recept**

BREER H LUEKEN W 1983 Transport of choline by membrane vesicles prepared from synaptosomes of insect nervous tissue. *Neurochem. Int. 5: 713-720* **TS ACh synaptosome**

BREER H MADLER U HEILIGENBERG H 1989 Immunocytochemical localization of glutamate decarboxylase in the mesothoracic ganglion of *Locusta migratoria. Tissue & Cell 21: 545-558* **TS glu thor anat enz**

BREER H SATTELLE DB 1987 Molecular properties and functions of insect acetylcholine receptors. *J. Insect Physiol. 33: 771-790* **TS ACh recept**

BREIDBACH O KUTSCH W 1990 Structural homology of identified motoneurones in larval and adult stages of hemi- and holometabolous insects. *J. comp. Neurol. 297: 392-409* **MN homol anat**

BROSEMER RW VOGELL W BÜCHER T 1963 Morphologische und enzymatische Muster bei der Entwicklung indirekter Flugmuskeln von *Locusta migratoria. Biochem. Z. 338: 854-910* **flight musc enz devel**

BROUSSE-GAURY P 1971 Déscription d'arcs reflexes neuro-endocriniens partant des ocelles chez quelques Orthoptères. *Bull. Biol. Fr. Belg. 105: 83-93* **oc NSC retroc anat brain**

BROWN MCS LUNT GG STAPLETON A 1987 (35S)-TBPS binding to ganglionic membranes of the locust (*Schistocerca gregaria*). *Biochem. Soc. Trans. 15: 503-504* **TS GABA**

BROWN MCS LUNT GG STAPLETON A 1989 Further characterization of the (35S)-TBPS binding site of the GABA receptor complex in locust (*Schistocerca gregaria*) ganglia membranes. *Comp. Biochem. Physiol. C 92: 9-13* **TS GABA**

BROWN RHJ 1967 Mechanism of locust jumping. *Nature 214: 939* **jump**

BRUCE M BUKOWNIK R ELDEFRAWI AT ELDEFRAWI ME GOODNOW R Jr KALLIMOPOULOS T KONNO K NAKANISHI K NIWA M USHERWOOD PNR 1990 Structure-activity relationships of analogues of the wasp toxin philanthotoxin: non-competitive antagonists of quisqualate receptors. *Toxicon 28: 1333-1346* **TS glu tox recept**

BRUNDELL P GOODNOW R KERRY CJ NAKANISHI K SUDAN HL USHERWOOD PNR 1991 Quisqualate-sensitive glutamate receptors of the locust *Schistocerca gregaria* are antagonised by intracellularly applied philanthotoxin and spermine. *Neurosci. Lett. 131: 196-200* **recept TS glu tox NM**

BRUNN DE BURROWS M WATSON AHD 1988 A central excitatory synapse between identified motor neurones in the locust. *J. Physiol. (Lond.) 407: 59P* **MN thor**

BUCHAN PB EVANS PD 1980 Use of an operational amplifier signal differentiator reveals that octopamine increases the rate of development of neurally evoked tension in insect muscle. *J. exp. Biol. 85: 349-352* **TS oct musc mod**

BUCHTHAL F WEIS-FOGH T ROSENFALCK P 1957 Twitch contractions of isolated flight muscle of locusts. *Acta physiol. Scand. 35: 246-276* **flight musc**

BUDD TC CLINTON P DELL A DUCE IR JOHNSON SJ QUICKE DLJ TAYLOR GW USHERWOOD PNR USOH G 1988 Isolation and characterisation of glutamate receptor antagonists from venoms of orb-web spiders. *Brain Res. 448: 30-39* **TS glu musc tox recept**

BUDD TC DUCE IR 1988 Immunocytochemical localisation of L-glutamate in motor innervation of locust (*Schistocerca gregaria*) skeletal muscle. *Pest. Sci. 23: 177-178* **musc anat glu**

BURNS MD 1973 The control of walking in Orthoptera. I. Leg movements in normal walking. *J. exp. Biol. 58: 45-58* **walk behav rom**

BURNS MD 1974 Structure and physiology of the locust femoral chordotonal organ. *J. Insect Physiol. 20: 1319-1339* **SO FCO anat propr**

BURNS MD USHERWOOD PNR 1978 Mechanical properties of locust extensor-tibiae muscles. *Comp. Biochem. Physiol. A 61: 85-95* **musc leg**

BURNS MD USHERWOOD PNR 1979 The control of walking in Orthoptera. II. Motor neurone activity in normal free-walking animals. *J. exp. Biol. 79: 69-98* **walk MN leg**

BURROWS M 1976 The influence of sensory inflow on the flight system of the locust. *In: Spencer Davies P (ed.), Perspectives in experimental biology. Vol.1: Zoology. Oxford & New York: Pergamon Press. 399-409* **MN SR flight**

BURROWS M 1973 Physiological and morphological properties of the metathoracic common inhibitory neuron of the locust. *J. comp. Physiol. 82: 59-78* **anat MN inhib**

BURROWS M 1973 The role of delayed excitation in the coordination of some metathoracic flight motoneurons in a locust. *J. comp. Physiol. 83: 135-164* **flight MN thor**

BURROWS M 1973 The morphology of an elevator and a depressor motoneuron of the hindwing of a locust. *J. comp. Physiol. 83: 165-178* **anat MN flight var thor**

BURROWS M 1974 Modes of activation of motoneurons controlling ventilatory movements of the locust abdomen. *Phil. Trans. R. Soc. Lond. (Biol.) 269: 29-48* **MN resp abd**

BURROWS M 1975 Integration by motoneurones in the central nervous system of insects. *In: Usherwood PNR Newth DR (eds.): "Simple" nervous systems. London: Edward Arnold. 345-379* **MN thor**

BURROWS M 1975 Monosynaptic connections between wing stretch receptors and flight motoneurones of the locust. *J. exp. Biol. 62: 189-220* **flight SO SR MN propr**

BURROWS M 1975 Coordinating interneurones of the locust which convey two patterns of motor commands: their connexions with ventilatory motoneurones. *J. exp. Biol. 63: 735-754* **resp MN IN thor Abd**

BURROWS M 1976 Neural control of flight in the locust. *In: Hermann RM Grillner S Stein PSG Stuart DG (eds.), Neural control of locomotion. Advances in behavioural biology No. 18. New York: Plenum Press. 419-438* **flight**

BURROWS M 1976 Flight mechanisms of the locust. *In: Hoyle G (ed.): Identified neurons and behaviour in Arthropods. New York: Plenum Press. 339-356* **flight jump aud teg thor propr**

BURROWS M 1978 Sources of variation in the output of locust spiracular motoneurones receiving common synaptic driving. *J. exp. Biol. 74: 175-186* **MN thor medN resp var**

BURROWS M 1978 Local interneurones and integration in locust ganglia. *Verh. dt. zool. Ges. 1978: 68-79* **IN MN leg thor nonspik rev**

BURROWS M 1979 Graded synaptic transmission between local premotor interneurons of the locust. *J. Neurophysiol. 42: 1108-1123* **nonspik IN MN thor local**

BURROWS M 1979 Synaptic potentials affect the release of transmitter from locust nonspiking interneurons. *Science 204: 81-83* **local IN TS nonspik**

BURROWS M 1980 The control of sets of motoneurons by local interneurones in the locust. *J. Physiol. (Paris) 298: 213-223* **nonspik leg MN anat**

BURROWS M 1980 The tracheal supply to the central nervous system of the locust. *Proc. R. Soc. Lond. (Biol.) 207: 63-78* **trach anat**

BURROWS M 1982 The physiology and morphology of median nerve motor neurones in the thoracic ganglion of the locust. *J. exp. Biol. 96: 325-342* **MN anat thor medN resp**

BURROWS M 1982 Interneurones coordinating the ventilatory movements of the thoracic spiracles in the locust. *J. exp. Biol. 97: 385-400* **IN resp thor**

BURROWS M 1985 Nonspiking and spiking local interneurones in the locust. *In: Selverston AI (ed.), Model neural networks and behavior. New York: Plenum Press. 109-125* **local SO IN thor rev nonspik**

BURROWS M 1985 The processing of mechanosensory information by spiking local interneurons in the locust. *J. Neurophysiol. 54: 463-478* **local SO IN thor**

BURROWS M 1987 Parallel processing of proprioceptive signals by spiking local interneurons and motor neurons in the locust. *J. Neurosci. 7: 1064-1080* **SO FCO local IN MN anat propr**

BURROWS M 1987 Inhibitory interactions between spiking and nonspiking local interneurones in the locust. *J. Neurosci. 7: 3282-3292* **IN thor local nonspik inhib**

BURROWS M 1987 Interactions between spiking and non-spiking local interneurones underlie local reflexes of locust legs. *J. Physiol. (Lond.) 390: 228P* **IN leg nonspik**

BURROWS M 1988 Responses of spiking local interneurones in the locust to proprioceptive signals from the femoral chordotonal organ. *J. comp. Physiol. A 164: 207-217* **SO FCO propr MN thor local IN**

BURROWS M 1989 Local circuits for the control of movement in insects. *In: Erber J Menzel Pflüger H-J Todt D (eds.), Proc. 2nd Internat. Congr. Neuroethology. Stuttgart & New York: Thieme Verlag. (NB - In Supplement!)* **leg thor IN SO reflex**

BURROWS M 1989 Effects of temperature on a central synapse between identified motor neurones in the locust. *J. comp. Physiol. A 165: 687-695* **MN temp thor**

BURROWS M 1989 Processing of mechanosensory signals in local reflex pathways of the locust. *J. exp. Biol. 146: 209-227* **thor IN nonspik leg SO mech rev**

BURROWS M HORRIDGE GA 1974 The organization of inputs to motoneurones of the locust metathoracic leg. *Phil. Trans. R. Soc. Lond. (Biol.) 269: 49-94* **MN leg IN thor**

BURROWS M HOYLE G 1973 Neural mechanisms underlying behaviour in the locust *Schistocerca gregaria*. III. Topography of limb motoneurons in the metathoracic ganglion. *J. Neurobiol. 4: 167-186* **anat MN leg thor**

BURROWS M LAURENT G 1989 Reflex circuits and the control of movement. *In: Durbin R Miall RC Mitchinson GM (eds.), The computing neuron. London: Addison-Wesley. 246-261* **IN MN leg thor nonspik**

BURROWS M LAURENT G FIELD LH 1988 Nonspiking local interneurones receive direct inputs from a proprioceptor and contribute to local reflexes of a locust hindleg. *J. Neurosci. 8: 3085-3093* **IN thor local propr SO FCO leg anat**

BURROWS M PFLÜGER H-J 1986 Processing by local interneurones of mechanosensory signals involved in a leg reflex of the locust. *J. Neurosci. 6: 2764-2777* **local IN SO leg anat mech**

BURROWS M PFLÜGER H-J 1988 Positive feedback loops from proprioceptors involved in leg movements of the locust. *J. comp. Physiol. A 163: 425-440* **SO propr MN IN nonspik thor kick innerv**

BURROWS M ROWELL CHF 1973 Connections between descending visual interneurons and metathoracic motoneurons in the locust. *J. comp. Physiol. 85: 221-234* **DN MN thor**

BURROWS M SIEGLER MVS 1976 Transmission without spikes between locust interneurones and motoneurones. *Nature 262: 222-224* **nonspik IN MN thor local anat**

BURROWS M SIEGLER MVS 1978 Graded synaptic transmission between local interneurones and motor neurons in the metathoracic ganglion of the locust. *J. Physiol. (Lond.) 265: 231-255* **local MN leg IN nonspik anat thor**

BURROWS M SIEGLER MVS 1982 Spiking local neurones mediate local reflexes. *Science 217: 650-652* **SO local MN leg anat thor**

BURROWS M SIEGLER MVS 1984 The morphological diversity and receptive fields of spiking local interneurones in the locust metathoracic ganglion. *J. comp. Neurol. 224: 483-508* **local anat SO mech thor**

BURROWS M SIEGLER MVS 1985 The organisation of receptive fields of spiking local interneurones in the locust with inputs from hair afferents. *J. Neurophysiol. 53: 1147-1157* **SO local thor mech**

BURROWS M WATKINS BL 1986 Spiking local interneurones in the mesothoracic ganglion of the locust: homologies with metathoracic interneurones. *J. comp. Neurol. 245: 29-40* **local MN leg var anat**

BURROWS M WATSON AHD BRUNN DE 1989 Physiological and ultrastructural characterization of a central synaptic connection between identified motor neurons in the locust. *Eur. J. Neurosci. 1: 111-126* **MN leg thor**

BURTT ET 1967 Some problems of insect vision. *In: Beament JWL Treherne JE (eds.), Insects and Physiology. Edinburgh and London: Oliver and Boyd. 217-232* **opt SO**

BURTT ET CATTON WT 1954 Visual perception of movement in the locust. *J. Physiol. (Lond.) 125: 566-580* **DN vis opt oc tech**

BURTT ET CATTON WT 1956 Electrical responses to visual stimulation in the optic lobes of the locust and certain other insects. *J. Physiol. (Lond.) 146: 492-514* **DN vis opt**

BURTT ET CATTON WT 1959 Transmission of visual responses in the nervous system of the locust. *J. Physiol. (Lond.) 146: 492-514* **DN vis opt**

BURTT ET CATTON WT 1959 Responses of single visual receptor units in the optic lobe of the locust. *J. Physiol. (Lond.) 148: 58-59* **opt**

BURTT ET CATTON WT 1959 Visual acuity of the compound eyes in three species of insects. *Proc. 15th. Int. Congr. Zool., London 1958: 535-536* **opt**

BURTT ET CATTON WT 1960 The properties of single unit discharges in the optic lobes of the locust. *J. Physiol. (Lond.) 154: 479-490* **opt**

BURTT ET CATTON WT 1961 Diffraction images in the compound eye. *J. Physiol. (Lond.) 159: 52P* **opt vis ret**

BURTT ET CATTON WT 1961 Visual acuity in insects. *J. Physiol. (Lond.) 159: 64P-66P* **opt vis**

BURTT ET CATTON WT 1962 Resolving power of the compound eye. *In: Beament JWL (ed.), Biological receptor mechanisms. Symp. Soc. Exp. Biol. 16. Cambridge: Cambridge University Press. 72-85* **opt vis**

BURTT ET CATTON WT 1962 A diffraction theory of insect vision. I. An experimental investigation of visual acuity and image formation in the compound eye of three species of insects. *Proc. R. Soc. Lond. (Biol.) 157: 53-82* **DN vis ret anat**

BURTT ET CATTON WT 1964 The potential profile of the insect compound eye and optic lobe. *J. Insect Physiol. 10: 689-710* **opt ret**

BURTT ET CATTON WT 1964 Potential changes in the eye and optic lobe of certain insects during light- and dark- adaptation. *J. Insect Physiol. 10: 865-886* **opt trach**

BURTT ET CATTON WT 1966 Perception by locusts of rotated patterns. *Science 151: 224* **DN vis**

BURTT ET CATTON WT 1969 Resolution of the locust eye measured by rotation of radial striped patterns. *Proc. R. Soc. Lond. (Biol.) 173: 513-529* **DN vis**

BURTT ET CATTON WT COSENS DJ 1963 Correlation between electrical potentials and visual threshold changes in an insect eye. *J. Physiol. (Lond.) 170: 8P* **vis opt ret**

BURTT ET CATTON WT COSENS DJ 1966 Electrical potential field in eye and optic lobe of locust: potential variations and change in visual threshold during light- and dark-adaptation. *In: Bernhard CG (ed.), The functional organization of the compound eye. London, Pergamon Press. 219-230* **opt**

BURTT ET CATTON WT 1966 Image formation and sensory transmission in the compound eye. *Adv. Insect Physiol. 13: 1-52* **vis ret**

BÜSCHGES A DRIESANG R PEARSON KG 1991 Projections and connections of forewing tegula afferences in the locust before and after the loss of the hindwing tegulae. *In: Elsner N Penzlin H (eds.), Synapse-transmission-modulation: Proc. 19th Göttingen Neurobiol. Conf. Stuttgart, New York: Thieme Verlag. Abstract 530* **flight propr regen IN teg**

BÜSCHGES A DRIESANG R RAMIREZ JM PEARSON KG 1991 Reorganization of sensory regulation in locust flight after partial deafferentation. *Soc. Neurosci. Abstr. 17: 1227* **flight propr regen IN teg**

BÜSCHGES A PEARSON KG 1990 Plasticity in the flight system of the locust: recovery of flight pattern after removal of wing proprioceptors. *In: Elsner N Roth G (eds.) Brain - Perception - Cognition: Proc. 18th Göttingen Neurobiol. Conf. Stuttgart, New York: Thieme Verlag. Abstract 69* **teg flight SO labil**

BUSCHGES A PEARSON KG 1990 Adaptive modification of the flight system of the locust after removal of wing proprioceptors. *J. exp. Biol. 157: 313-333* **flight propr behav**

BÜSCHGES A PEARSON KG 1990 Plasticity in the flight system of the locust. *Soc. Neurosci. Abstr. 16: 1225* **flight propr behav**

BÜSCHGES A PEARSON KG 1991 Adaptive modifications in the flight system of the locust after the removal of wing proprioceptors. *J. exp. Biol. 157: 313-334* **flight SO propr teg**

BÜSCHGES A RAMIREZ JM PEARSON KG 1991 Reorganization of afferent pathways following partial deafferentation of the flight system of the locust. *In: Elsner N Penzlin H (eds.), Synapse-transmission-modulation: Proc. 19th Göttingen Neurobiol. Conf. Stuttgart, New York: Thieme Verlag. Abstract 531* **flight propr regen IN**

BUSNEL MC BURKHARDT D 1962 An electrophysiological study of the phonokinetic reaction in *Locusta migratoria migratorioides* (L.). *Symp. Zool. Soc. Lond. 7: 13-44* **aud SO behav**

BUSNEL MC BUSNEL RG 1956 Sur une phonocinèse de certains acridiens à des signaux acoustiques synthétiques. *C. R. Acad. Sci. Paris Ser. D 242: 292-293* **aud behav**

BUTLIN RK HEWITT GM 1986 The response of female grasshoppers to male song. *Anim. Behav. 34: 1896-1899* **strid behav**

BYCROFT BW BLAGBROUGH IS USHERWOOD PNR 1990 Natural and synthetic polyamine derivatives as antagonists of glutamate receptors: an emerging structure/activity profile. *Eur. J. Pharmacol. 183: 578-579* **TS glu tox recept**

CAMHI JM 1969 Locust wind receptors I. Transducer mechanisms and sensory response. *J. exp. Biol. 50: 335-348* **flight windh SO mech**

CAMHI JM 1969 Locust wind receptors II. Interneurones in the cervical connective. *J. exp. Biol. 50: 349-369* **flight windh SO mech**

CAMHI JM 1969 Locust wind receptors. III. Contribution to flight initiation and lift control. *J. exp. Biol. 50: 363-374* **flight windh SO mech**

CAMHI JM 1970 Yaw correcting postural changes in locusts. *J. exp. Biol. 52: 519-532* **flight windh SO mech behav**

CAMHI JM 1970 Sensory control of abdomen posture in flying locusts. *J. exp. Biol. 52: 533-538* **flight windh SO mech vis DN**

CAMHI JM 1971 Flight orientation in locusts. *Sci. Am. 225: 74-81* **flight windh SO mech**

CAMHI JM 1974 Neural mechanisms of response modification in insects. *In: Barton Browne L (ed.), Experimental analysis of insect behaviour. Berlin, Heidelberg, New York: Springer Verlag. 60-85* **labil flight**

CAMHI JM HINKLE M 1972 Attentiveness to sensory stimuli: central control in locusts. *Science 175: 550-552* **flight windh SO mech**

CAMHI JM HINKLE M 1974 Response modification by central flight oscillator of locusts. *J. exp. Biol. 60: 477-492* **flight windh SO mech resp rhythm anat**

CAMIER M GIRARDIE J REMY C GIRARDIE A COHEN P 1980 Identification of immunoreactive neurophysin-like proteins in the central nervous system of an insect, *Locusta migratoria. Biochem. Biophys. Res. Comm. 93: 792-796* **brain retroc NSC pep neurophys**

CAMPBELL JI 1961 The anatomy of the mesothorax of *Locusta migratoria migratorioides* R. & F. *Proc. zool. Soc. Lond. 137: 403-432* **anat thor**

CANDY DJ 1968 Carbohydrate utilization by locust thoracic muscle in vitro. *Biochem. J. 108: 12P* **metab musc**

CANDY DJ 1970 Metabolic studies on locust flight muscle using a new perfusion technique. *J. Insect Physiol. 16: 531-543* **metab musc flight**

CANDY DJ 1974 The control of muscle trehalase activity during locust flight. *Biochem. Soc. Trans. 2: 1107-1109* **metab musc flight**

CANDY DJ 1978 The regulation of locust flight muscle metabolism by octopamine and other compounds. *Insect Biochem. 8: 177-181* **TS oct musc flight mod**

CANDY DJ HALL LM SPENCER IM 1976 The metabolism of glycerol in the locust *Schistocerca gregaria* during flight. *J. Insect Physiol. 22: 583-587* **metab flight**

CARLSEN J HERMAN WS CHRISTENSEN M JOSEFSSON L 1979 Characterization of a second peptide with adipokinetic and red-pigment concentrating activty from locust corpora cardiaca. *Insect Biochem. 9: 497-501* **pep retroc AKH**

CARLSON JG 1938 Mitotic behavior of induced chromosomal fragments lacking spindle attachments in the neuroblasts of the grasshopper *Proc. nat. Acad. Sci. USA 24: 500-507* **devel genet**

CARLSON JG 1938 Some effects of X-radiation on the neuroblast chromosomes of the grasshopper, *Chortophaga viridifasciata. Genetics 23: 596-609* **devel genet tox**

CARLSON JG 1940 Immediate effects of 25O r of X-rays on the different stages of mitosis in neuroblasts of the grasshopper, *Chortophaga viridifasciata. J. Morphol. 66: 11-23* **devel genet tox**

CARLSON JG 1941 An analysis of X-ray induced single breaks in neuroblast chromosomes of the grasshopper, *Chortophaga viridifasciata Proc. nat. Acad. Sci. USA 27: 42-47* **devel genet tox**

CARLSON JG 1946 Protoplasmic viscosity changes in different regions of the grasshopper neuroblast during mitosis *Biol. Bull. Woods Hole 90: 109-121* **devel genet**

CARLSON JG 1952 Microdissection studies of the dividing neuroblast of the grasshopper, *Chortophaga viridifasciata* (deGeer). *Chromosoma 5: 199-22O* **devel genet**

CARLSON JG 1959 Embryological development of neuroblasts. *In: Campbell FL (ed.), Physiology of insect development. Chicago: University of Chicago Press. 40-47* **devel genet**

CARLSON JG 1961 The grasshopper neuroblast culture technique and its value in radiobiological studies. *Ann. N.Y. Acad. Sci. 95: 932-941* **devel genet tox cellCult**

CARLSON JG 1969 A detailed analysis of X-ray-induced prophase delay and reversion of grasshopper neuroblasts in culture. *Radiation Res. 37: 1-14* **devel genet tox**

CARLSON JG 1976 Mitotic effect of monochromatic UV-radiation at 225, 265 and 280 nanometers on 11 stages of the cell cycle of the grasshopper neuroblast in culture. Part 1. Overall retardation from the stage irradiated to nuclear membrane breakdown. *Radiat. Res. 68: 57-74* **devel genet tox**

CARLSON JG 1976 Mitotic effect of monochromatic UV-radiation at 225, 265 and 280 nanometers on 11 stages of the cell cycle of the grasshopper neuroblast in culture. Part 2. Changes in progression rate between the stage irradiated and nuclear membrane breakdown. *Radiat. Res. 68: 75-83* **devel genet tox**

CARLSON JG 1977 Anaphase chromosome movement in the unequally dividing grasshopper neuroblast and its relation to other cells *Chromosoma 64: 191-206* **devel genet**

CARLSON JG GAULDEN ME 1951 Cytological effects of colchicine on the grasshopper neuroblast in vitro with special reference to the origin of the spindle *Exptl. Cell Res. 2: 416-433.* **devel genet tox**

CARLSON JG GAULDEN ME 1964 Grasshopper neuroblast techniques. *In: Prescott DM (ed.), Methods of cell physiology. New York: Academic Press. 229-276* **devel genet**

CARLSON JG HARRINGTON NG GAULDEN ME 1953 Mitotic effects of prolonged irradiation with low-intensity gamma rays on the *Chortophaga* neuroblast. *Biol. Bull. Woods Hole 104: 313-322* **devel genet tox**

CARLSON JG HOLLAENDER A 1944 Immediate effects of low doses of ultraviolet radiation of wavelength 2537 Å on mitosis in the grasshopper neuroblast *J. cell. comp. Physiol. 23: 157-169* **devel genet tox**

CARLSON JG HOLLAENDER A 1945 Intensity effects of ultraviolet radiation of wavelength 2537 Å on mitosis in the grasshopper neuroblast. *J. cell. comp. Physiol. 26: 165-173* **devel genet tox**

CARLSON JG McGRATH RA LEACH WM 1963 A squash technique for autoradiography of grasshopper neuroblasts. *J. R. micr. Soc. Lond. 82: 55-56* **devel genet tech**

CARLSON JG McMASTER RD 1951 Nucleolar changes induced in the grasshopper neuroblast by different wavelengths of ultraviolet radiation and their capacity for photorecovery *Exptl. Cell Res. 2: 434-444* **devel genet tox**

CARLSON JG SNYDER ML HOLLAENDER A 1949 Relation of gamma-ray dosage rate to mitotic effect in the grasshopper neuroblast *J. cell. comp. Physiology 33: 365-372* **devel genet tox**

CARLSON JG STEPHENS RE 1972 Action of actinomycin D on grasshopper neuroblasts in culture *Exptl. Cell Res. 46: 42-50* **devel cellCult tox**

CARLSON JG STEPHENS RE 1974 Mitotic acceleration induced in grasshopper neuroblasts by small doses of monochromatic radiation. *Photochem. Photobiol. 20: 1-7* **devel genet tox**

CARPENTER EM BASTIANI MJ 1989 Developmental expression of a glial antigen in grasshopper embryos. *Soc. Neurosci. Abstr. 15: 649* **devel glia**

CARPENTER EM BASTIANI MJ 1990 Expression and characterization of nervous system antigens in grasshopper embryos. *Soc. Neurosci. Abstr. 16: 303* **tech devel**

CARPENTER EM BASTIANI MJ 1991 Developmental expression of REGA-1, a regionally expressed glial antigen in the central nervous system of grasshopper embryos. *J. Neurosci. 11: 277-286* **devel glia**

CASSIER P 1962 Rôle des ocelles frontaux chez *Locusta migratoria migratorioides* (R. et F.) *Insects Sociaux 9: 213* **oc behav**

CASSIER P 1965 Contribution a l'étude des réactions photomenotactiques du criquet migrateur *Locusta migratoria migratorioides* (R. et F.): leur determinisme sensoriel. *Insects Sociaux 12: 365-381* **oc behav**

CASSIER P FAIN-MAUREL M-A 1970 Contribution à la étude infrastructurale du système neurosécréteur rétrocérébrale chez *Locusta migratoria migratorioides* (R. et F.). I. Les corpora cardiaca. *Z. Zellforsch. mikr. Anat. 111: 471-482* **retroc anat NSC CC fineStruc**

CASSIER P FAIN-MAUREL M-A 1970 Contribution à la étude infrastructurale du système neurosécréteur rétrocérébrale chez *Locusta migratoria migratorioides* (R. et F.). II. Le transit des neurosécretions. *Z. Zellforsch. mikr. Anat. 111: 483-492* **retroc anat NSC CC fineStruc**

CATTON WT 1976 Peaking of response in locust visual interneurones to objects of small angular subtense. *J. Insect Physiol. 22: 1475-1484.* **DN vis**

CATTON WT 1978 Responses of locust visual interneurones to transiently-illuminated slits, paired spots, annuli and disc arrays. *J. Insect Physiol. 24: 285-292* **DN vis**

CATTON WT 1978 Spatial frequency peaks in the responses of locust DCMD axon to rotation of periodic patterns, lines and dot arrays. *J. Insect Physiol. 24: 355-362* **DN vis**

CATTON WT 1980 Tonic effects of stationary luminous slits on the discharge rates of some locust visual interneurones. *J. Insect Physiol. 26: 373-379* **vis DN**

CATTON WT 1982 The effects of stimulus area and intensity on the on/off ratio of some locust visual interneurones. *J. Insect Physiol. 28: 285-291* **vis DN**

CATTON WT 1985 Time-course and spatial spread of inhibition in the locust compound eye. *Zool. Anz. 215: 201-210* **vis ret**

CATTON WT 1988 Discharge patterns of locust visual interneurones. *Physiol. Entomol. 13: 147-152* **vis IN**

CATTON WT CHAKRABORTY A 1969 Single neuron responses to visual and mechanical stimuli in the thoracic nerve cord of the locust. *J. Insect Physiol. 15: 245-258* **DN VNC vis wind aud IN oc**

CAUDY M BENTLEY D 1986 Pioneer growth cone steering along a series of neuronal and non-neuronal cues of different affinities. *J. Neurosci. 6: 1781-1795* **devel leg**

CAUDY M BENTLEY D 1986 Pioneer growth cone morphologies reveal proximal increases in substrate affinity within leg segments of grasshopper embryos. *J. Neurosci. 6: 364-379* **devel leg**

CAZAL M 1969 Action des corpora cardiaca sur le péristaltisme intestinal de *Locusta migratoria. Archs. Zool. exp. gén. 110: 83-90* **retroc CC gut**

CAZAL M 1971 Action des corpora cardiaca sur la tréhalosémie et la glycémie de *Locusta migratoria* L. *C. R. hebd. Seanc. Acad. Sci. Paris Ser D 272: 2596-2599* **retroc CC metab**

CAZAL M GIRARDIE A 1968 Contrôle humoral de l'équilibre hydrique chez *Locusta migratoria migratorioides. J. Insect Physiol. 14: 655-668.* **ADH**

CAZAL M GIRARDIE A BENTZ F 1971 Action des corpora cardiaca sur le développement génital de *Locusta migratoria. Archs. Zool. exp. gén. 112: 292-300* **retroc CC genit**

CAZAL P 1948 Les glandes endocrines rétro-cérébrales des Insectes. (Etude morphologique). *Bull. Biol. Fr. Belg. Suppl. 32. 227 pp.* **anat retroc**

CAZAL P GUERRIER Y 1946 Recherches sur les glandes endocrines rétro-cérébrales des Insects. I. Etude morphologique chez les Orthoptères. *Archs. Zool. exp. gén. 84: 303-34* **retroc anat**

CERF JA GRUNDFEST H HOYLE G MC CANN FV 1959 The mechanism of dual responsiveness in muscle fibers of the grasshopper *Romalea microptera. J. gen. Physiol. 43: 377-395* **musc rom**

CHALAYE D 1965 Neurosécrétion au niveau de la chaîne nerveuse de *Locusta migratoria migratorioides* R. et F. (Orthoptère Acridien). *Bull. Soc. Zool. Fr. 92: 87-108* **NSC thor abd anat**

CHALAYE D 1966 Recherches sur la destination des produits de neurosécrétion de la chaîne nerveuse ventrale du criquet migrateur, *Locusta migratoria. C. R. Acad. Sci. Paris Ser. D 262: 161-164* **pep NSC thor**

CHALAYE D 1969 La tréhalosémie et son contrôle neuroendocrine chez le criquet migrateur *Locusta migratoria migratorioides.* 2. Rôle de corpora cardiaca et des organes perisympathiques. *C. R. hebd. Seanc. Acad. Sci. Paris Ser D 268: 3111-3114* **retroc CC metab**

CHALAYE D 1974 Ultrastructure de la masse ganglionnaire metathoracique de *Locusta migratoria migratorioides* (R. et F.) (Orthoptère). I. Les cellules neurosécrétrices et leurs prolongements dans le neuropile. *Acrida 3: 19-33* **NSC thor abd fineStruc**

CHALAYE D 1974 Ultrastructure de la masse ganglionnaire metathoracique de *Locusta migratoria migratorioides* (R. et F.) (Orthoptère). II. Les organes périsympathiques abdominaux et thoraciques. *Acrida 3: 35-46* **NSC thor abd fineStruc**

CHALAYE D 1975 Etude ultrastructurale de cellules neurosécrétrices du ganglion sous-oesophagien de *Locusta migratoria migratorioides* (R. & F.) (Orthoptères). *Acrida 4: 19-32* **fineStruc NSC suboes**

CHANG S HO RK GOODMAN CS 1983 Selective groups of neuronal and mesodermal cells recognized early in grasshopper embryogenesis by a monoclonal antibody. *Dev. Brain Res. 9: 297-304* **devel tech**

CHAPMAN RF 1958 A field study of the potassium concentration in the blood of the red locust *Nomadacris septemfasciata* (Serv.), in relation to its activity. *Anim. Behav. 6: 60-67* **blood behav**

CHAPMAN RF 1966 The mouthparts of *Xenocheila zarudnyi* (Orthoptera, Acrididae). *J. Zool. (Lond.) 143: 277-288* **anat mpt**

CHAPMAN RF 1970 Growth of the embryo of *Ornithacris turbida* (Walker) (Orthoptera: Acrididae) with special reference to the antennae. *Proc. R. entomol. Soc Lond. (A) 45: 29-38* **ant SO devel**

CHAPMAN RF 1974 The control of feeding in acridids. *Acrida 3: xxxix-xlviii* **feed SO behav**

CHAPMAN RF 1982 Chemoreception: the significance of receptor numbers. *Adv. Insect Physiol. 16: 247-355* **SO mpt feed chem**

CHAPMAN RF 1988 Sensory aspects of host-plant recognition by Acridoidea: questions associated with the multiplicity of receptors and variability of response. *J. Insect Physiol. 34: 167-74* **rev SO feed mpt chem**

CHAPMAN RF 1990 Food selection. *In: Chapman RF Joern A (eds.), Biology of grasshoppers. New York: J. Wiley & Sons Inc. 39-72* **feed behav SO chem rev**

CHAPMAN RF ASCOLI-CHRISTENSEN A WHITE PR 1991 Sensory coding for feeding deterrence in the grasshopper *Schistocerca americana. J. exp. Biol. 158: 241-260* **SO chem feed leg**

CHAPMAN RF BEERLING EAM 1990 The pattern of feeding of first instar nymphs of *Schistocerca americana. Physiol. Entomol. 15: 1-12* **feed behav**

CHAPMAN RF BERNAYS EA 1989 Insect behaviour at the plant surface and learning as aspects of host plant selection. *Experientia 45: 215-222* **feed behav learn**

CHAPMAN RF BERNAYS EA WYATT T 1988 Chemical aspects of host-plant specificity in three *Larrea*-feeding grasshoppers. *J. chem. Ecol. 14: 557-575* **feed behav chem mpt**

CHAPMAN RF BLANEY WM 1979 How animals perceive secondary compounds. *In: Rosenthal GA Janzen DJ (eds.), Herbivores: their interaction with secondary plant metabolites. New York: Academic Press. 161-198* **mpt SO chem behav**

CHAPMAN RF FRASER J 1989 The chemosensory system of the monophagous grasshopper, *Bootettix argentatus* Bruner (Orthoptera: Acrididae). *Int. J. Insect Morphol. Embryol. 18: 111- 118* **SO feed mpt**

CHAPMAN RF GREENWOOD M 1986 Changes in distribution and abundance of antennal sensilla during growth of *Locusta migratoria* L. (Orthoptera: Acrididae). *Int. J. Insect Morphol. Embryol. 15: 83-96* **ant SO mpt**

CHAPMAN RF PAGE WW BERNAYS EA 1981 A novel method of sound production by an acridoid, *Zonocerus variegatus* (L.) (Orthoptera: Pyrgomorphidae). *Acrida 10: 51-60* **flight musc strid anat pyrg**

CHAPMAN RF THOMAS JG 1978 The numbers and distribution of sensilla on the mouthparts of Acridoidea. *Acrida 7: 115-148* **SO mpt feed anat**

CHARLET M GOLTZENE F HOFFMANN JA 1979 Experimental evidence for a neuroendocrine control of ecdysone biosynthesis in adult females of *Locusta migratoria. J. Insect Physiol. 25: 463-466* **pep retroc NSC brain**

CHAUVIN R 1938 Anatomie et histologie du tube digestif de *Schistocerca gregaria. Bull. Soc. Hist. Nat. Afr. N. 28: 488-499* **morph gut**

CHEESEMAN P GOLDSWORTHY GJ 1979 The release of adipokinetic hormone during flight and starvation in *Locusta. Gen. comp. Endocrinol. 37: 35-43* **pep retroc flight AKH**

CHEESEMAN P JUTSUM AR GOLDSWORTHY GJ 1976 Quantitative studies on the release of locust adipokinetic hormone. *Physiol. Entomol. 1: 115-121* **pep retroc flight AKH**

CHÉZE G 1968 Contribution a l'étude des fibres nerveuses géantes de *Locusta migratoria migratorioides. Ann. Soc. entomol. Fr. N.S. 4: 207-211* **IN GF**

CHINO H DOWNER RGH TAKAHASHI 1986 Effect of adipokinetic hormone on the structure and properties of lipophorin in insect haemolymph. *J. Lipid Res. 27: 21-29* **pep AKH blood**

CHUDAKOVA IV 1960 [Analysis of the after-effect mechanism of the neuromuscular apparatus in insects.] (in Russian) *J. Physiol. USSR 46: 1044-1049* **musc thor**

CHUDAKOVA IV 1960 [On after-discharge in the neuromuscular apparatus of insects]. (in Russian) *Zhuzn. obsch Biol: 77-80* **musc thor**

CHUDAKOVA IV 1962 [Variation in the excitability of motoneurons of the thoracic ganglion in *Locusta migratoria* L. in response to the stimulation of the supraoesophageal ganglion.] (in Russian) *Dok. Akad. Nauk. SSSR 143: 743-745* **MN thor brainStim**

CHUDAKOVA IV 1963 [Nervous regulation of muscle activity of insects (An investigation on the asiatic locust).] (in Russian) *Dissertation, Moscow, USSR* **musc flight thor MN**

CHUDAKOVA IV 1963 [Postembryonic variations in the functional properties of wing muscles in asiatic locusts.] (in Russian) *Dok. Akad. Nauk. SSSR 152: 1490-1493* **musc flight thor**

CHYB S SIMPSON SJ 1990 Dietary selection in adult *Locusta migratoria* L. *Entomol. exp. appl. 56: 47-60* **feed behav**

CLARK RB DONALDSON PL GRATION KAF LAMBERT JJ PIEK T SPANJER W USHERWOOD PNR 1980 Post-synaptic block at neuromuscular junctions on locust muscle by o-philanthotoxin. *J. Physiol. (Lond.) 310: 8P* **TS glu NM tox**

CLARK RB GRATION KAF USHERWOOD PNR 1977 Glutamate D-receptors on the locust retractor unguis muscle. *J. Physiol. (Lond.) 276: 3-4P* **TS glu NM recept**

CLARK RB GRATION KAF USHERWOOD PNR 1979 Responses to DL-ibotenic acid at locust glutamatergic neuromuscular junctions. *Br. J. Pharmacol. 66: 267-273* **TS glu NM tox**

CLARK RB GRATION KAF USHERWOOD PNR 1979 Desensitization of glutamate receptors on innervated and denervated locust muscle fibres. *J. Physiol. (Lond.) 290: 551-568* **TS glu NM desens recept**

CLARK RB GRATION KAF USHERWOOD PNR 1979 Agonist-dependent desensitization of glutamate receptors at locust nerve-muscle junctions. *J. Physiol. (Lond.) 295: 93-94* **TS glu NM desens recept**

CLARK RB GRATION KAF USHERWOOD PNR 1979 Ion channels opened by glutamate agonists on locust muscle membrane. *J. Physiol. (Lond.) 301: 60-61* **musc glu channel TS**

CLARK RB GRATION KAF USHERWOOD PNR 1979 Relative "trophic" influences of excitatory and inhibitory innervation of locust skeletal muscle fibres. *Nature 280: 679-682* **NM musc**

CLARK RB GRATION KAF USHERWOOD PNR 1980 Influence of glutamate and aspartate on time course of decay of excitatory synaptic currents at locust neuromuscular junctions. *Brain Res. 192: 205-216* **TS glu NM asp**

CLARK RB GRATION KAF USHERWOOD PNR 1982 Influence of sodium and calcium ions and membrane potential on glutamate receptor desensitization. *Comp. Biochem. Physiol. C 72: 1-7* **TS glu musc desens recept**

CLARKE KU GRENVILLE H 1960 Nervous control of movements in the foregut of *Schistocerca gregaria* Forsk. *Nature 186: 98-99* **gut rhythm SGG innerv**

CLARKE KU LANGLEY PA 1963 Studies on the initiation of growth and moulting in *Locusta migratoria migratorioides* R. & F. I. The time and nature of the initiating stimulus. *J. Insect Physiol. 9: 287-292* **NSC SGG moult**

CLARKE KU LANGLEY PA 1963 Studies on the initiation of growth and moulting in *Locusta migratoria migratorioides* R. & F. II. The role of the stomatogastric nervous system. *J. Insect Physiol. 9: 363-373* **NSC SGG moult**

CLARKE KU LANGLEY PA 1963 Studies on the initiation of growth and moulting in *Locusta migratoria migratorioides* R. & F. III. The role of the frontal ganglion. *J. Insect Physiol. 9: 411-422* **NSC SGG moult**

CLARKE KU LANGLEY PA 1963 Studies on the initiation of growth and moulting in *Locusta migratoria migratorioides* R. & F. IV. The relationship between the stomatogastric nervous system and neurosecretion. *J. Insect Physiol. 9: 423-430* **NSC SGG moult**

CLEMENT EM STRANG RHC 1978 A comparison of some aspects of the physiology and metabolism of the nervous system of the locust *Schistocerca gregaria* in vitro with those in vivo. *J. Neurochem. 31: 135-145* **metab**

CLEMENTS AN MAY TE 1974 Studies on locust neuromuscular physiology in relation to glutamic acid. *J. exp. Biol. 60: 673-705* **NM glu TS**

CLEMENTS AN MAY TE 1974 Pharmacological studies on a locust neuromuscular preparation. *J. exp. Biol. 61: 421-442* **musc NM TS glu asp**

CLOAREC A GOURANTON J 1965 Contribution à l'étude de l'organogenèse des centres nerveux protocèrèbraux du criquet *Locusta migratoria migratorioides* R. & F. *Bull. Biol. F-r. Belg. 99: 357-368* **anat devel brain**

CLOUPEAU M DEVILLIERS JF DEVEZEAUX D 1979 Direct measurements of instantaneous lift in desert locust: comparison with Jensen's experiments on detached wings. *J. exp. Biol. 80: 1-15* **flight aero**

COCHRANE DG ELDER HY USHERWOOD PNR 1968 Electrical, mechanical and ultrastructural properties of tonic and phasic muscle fibres in the locust (*Schistocerca gregaria*). *J. Physiol. (Lond.) 200: 68-69P* **musc fineStruc**

COCHRANE DG ELDER HY USHERWOOD PNR 1972 Physiology and ultrastructure of phasic and tonic skeletal muscle fibres in the locust, *Schistocerca gregaria. J. Cell Sci. 10: 419-441* **musc fineStruc**

COILLOT JP 1975 La conversion analogique-numerique des récepteurs à l'étirement de la patte metathoracique du criquet *Schistocerca gregaria. J. Insect Physiol. 21: 423-433* **propr leg SO**

COILLOT JP BOISTEL J 1968 Localisation et description de récepteurs à l'étirement au niveau de l'articulation tibio-femorale de la patte sauteuse du criquet, *Schistocerca gregaria. J. Insect Physiol. 14: 1661-1667* **propr SO leg anat**

COILLOT JP BOISTEL J 1969 Etude de l'activité électrique propagée de récepteurs à l'étirement de la patte metathoracique du criquet, *Schistocerca gregaria. J. Insect Physiol. 15: 1449-1470* **propr SO leg**

COKL A KALMRING K WITTIG H 1977 The responses of auditory ventral cord neurons of *Locusta migratoria* to vibration stimuli. *J. comp. Physiol. A 120: 161-172* **aud vibr subg SO IN thor**

COKL A OTTO C KALMRING K 1985 The processing of directional vibratory signals in the ventral nerve cord of *Locusta migratoria. J. comp. Physiol. A 156: 45-52* **aud vibr IN thor anat**

COLLETT TS 1978 Peering - a locust behaviour pattern for obtaining parallax information. *J. exp. Biol. 76: 237-241* **vis OM head neck**

COLLETT TS PATERSON CJ 1991 Relative motion parallax and target localisation in the locust, *Schistocerca gregaria . J. comp. Physiol. A 169: 615-622* **vis behav jump**

CONDIC ML BENTLEY D 1989 Removal of the basal lamina in vivo reveals growth cone-basal lamina adhesive interaction and axonal tension in grasshopper embryos. *J. Neurosci. 9: 2678-2686* **devel ECM**

CONDIC ML BENTLEY D 1989 Pioneer growth cone adhesion in vivo to boundary cells and neurons after enzymatic removal of basic lamina in grasshopper embryos. *J. Neurosci. 9: 2687-2696* **devel ECM**

CONDIC ML BENTLEY D 1989 Pioneer neuron pathfinding from normal and ectopic locations in vivo after removal of the basal lamina. *Neuron 3: 427-439* **devel ECM**

CONDIC ML LEFCORT F BENTLEY D 1989 Selective recognition between embryonic afferent neurons of grasshopper appendages in vitro. *Dev. Biol. 135: 221-230* **devel**

COOK AG 1972 The ultrastructure of the A1 sensilla on the posterior surface of the clypeo-labrum of *Locusta migratoria migratorioides* (R. & F.). *Z. Zellforsch. mikr. Anat. 134: 539-554* **SO mpt fineStruc**

COOK AG 1976 The structure and function of the clypeo-labrum of *Locusta migratoria* (L.) (Orthoptera: Acrididae). *Dissertation, University of London, UK.* **anat mpt SO feed**

COOK AG 1977 The anatomy of the clypeo-labrum of *Locusta migratoria* (L.) (Orthoptera: Acrididae). *Acrida 6: 287-306* **anat mpt SO feed rev**

COOK AG 1979 The ultrastructure of some sensilla on the clypeo-labrum of *Locusta migratoria* (L.) (Orthoptera: Acrididae). *Acrida 8: 47-62* **fineStruc mpt SO feed**

COOK J RAMSEY RL USHERWOOD PNR 1985 Giga-ohm recordings of glutamate receptor-gated channels from locust muscle in vitro. *J. Physiol. (Lond.) 365: 16P* **glu musc TS channel recept**

COOK PM 1951 Observations on giant fibres of the nervous system of *Locusta migratoria. Quart. J. micr. Sci. 92: 297-305* **IN GF abd**

COOTER RJ 1979 Visually induced yaw movements in the flying locust *Schistocerca gregaria* (Forsk.). *J. comp. Physiol. A 131: 67-78* **flight vis aero**

COOTER RJ 1973 Flight and landing posture in locust hoppers. *Acrida 2: 307-318* **flight devel behav**

COOTER RJ 1989 Swarm flight behaviour in flies and locusts. *In: Goldsworthy GJ Wheeler CH (eds.), Insect flight. Boca Raton, Florida: CRC. Pp 165-203* **flight behav dipt**

COOTER RJ BAKER PS 1977 Weis-Fogh clap and fling mechanism in *Locusta. Nature 269: 53-54* **flight aero**

CORNWELL PD 1955 The functions of the ocelli of *Calliphora* (Diptera) and *Locusta* (Orthoptera). *J. exp. Biol. 32: 217-233* **oc dipt**

CORRETTE B GEWECKE M 1983 Die Aktivität der Antennenbahn und intrazellulär abgeleiteter Neuronen des Unterschlundganglions bei fixiert fliegenden Heuschrecken (*Locusta migratoria* L.). *Verh. dt. zool. Ges. 1983: 275* **flight ant IN suboes**

COSENS DJ 1966 Visual sensitivity in the light- and dark-adapted eye of the desert locust. *J. Insect Physiol. 12: 871-890* **vis**

COSENS DJ 1967 Extracellular potentials in the locust eye and optic lobe. *J. Insect Physiol. 13: 1373-1386* **opt ret**

COUILLAUD F GIRARDIE A GIRARDIE J 1990 Identification of gonadotropic and antigonadotropic factors from the nervous part of the corpora cardiaca in the african locust. *Int. J. Invert. Reprod. Devel. 16: 17-22* **genit hormX retroc CC**

COUILLAUD F GIRARDIE A 1990 Disconnected corpora allata provide a novel assay for allatostimulating factors. *J. exp. Zool. 253: 3320-3333* **retroc CAll**

COUILLAUD F MAUCHAMP B GIRARDIE A 1987 Biological, radiochemical and physicochemical evidence for the low activity of disconnected corpora allata in locust. *J. Insect Physiol. 33: 223-228* **retroc CAll**

CRESCITELLI F JAHN TL 1939 The electrical response of the dark-adapted grasshopper eye to various intensities of illumination and to different qualities of light. *J. cell. comp. Physiol. 13: 105-112* **ret**

CRESCITELLI F JAHN TL 1939 The effect of temperature on the electrical response of the grasshopper eye. *J. cell. comp. Physiol. 14: 13-27* **ret temp**

CRESCITELLI F JAHN TL 1942 Oscillatory electrical activity in the insect compound eye. *J. cell. comp. Physiol. 19: 47-56* **opt**

CROSSMAN AR KERKUT GA PITMAN RM WALKER RJ 1971 Electrically excitable nerve cell bodies in the central ganglia of two insect species, *Periplaneta americana* and *Schistocerca gregaria.* Investigation of cell geometry and morphology by intracellular dye injection. *Comp. Biochem. Physiol. A 40: 579-594* **DUM anat thor dict**

CULL-CANDY SG MILEDI R PARKER I 1981 Single glutamate-activated channels recorded from locust muscle fibres with perfused patch electrodes. *J. Physiol. (Lond.) 321: 195-210* **musc TS glu NM channel**

CULL-CANDY SG NEAL H USHERWOOD PNR 1973 Action of black widow spider venom on an aminergic synapse. *Nature 241: 353-354* **TS glu NM tox**

CULL-CANDY SG PARKER I 1981 Rapid kinetics of single glutamate-receptor channels. *Nature 295: 410-412* **musc TS glu NM channel recept**

CULL-CANDY SG USHERWOOD PNR 1973 Two populations of L-glutamate receptors on locust muscle fibres. *Nature New Biol. 246: 62-64* **musc TS glu NM recept**

CULL-CANDY SG 1982 Properties of postsynaptic channels activated by glutamate and GABA in locust muscle fibres. *In: Neuropharmacology of insects. Pitman, London (Ciba Foundation Symp. 88). Pp 70-87* **rev musc TS glu NM channel**

CUPO A PROUX J 1982 Détection et quantification des précurseurs de la "vasopressin-like" peptide chez le criquet migrateur. *Bull. Soc. Neuroendocr. 1: 20-21* **pep DH**

CUPO A PROUX J 1983 Biochemical characterization of a vasopressin-like neuropeptide in *Locusta migratoria.* Evidence of a high molecular weight protein encoding vasopressin sequence. *Neuropeptides 3: 309-318* **pep DH brain**

CUPO A PROUX J 1983 Biochemical and biosynthetic characterization of the locust vasopressin-like hormone. *Neuropeptides 4: 309-312* **pep DH brain**

CUTHBERT BA EVANS PD 1989 A comparison of the effects of FMRFamide-like peptides on locust heart and skeletal muscle. *J. exp. Biol. 144: 395-415* **pep FMRF heart musc**

DALE JF TOBE SS 1990 The endocrine basis of locust phase polymorphism. *In: Chapman RF Joern A (eds.), Biology of grasshoppers. New York: J. Wiley & Sons Inc. 393-413* **hormX phase NSC rev**

DANDO J CHANUSSOT B DANDO MR 1968 Le système nerveuse stomodéal postcéphalique de *Schistocerca gregaria* Forskål (Orthoptère) et *Blabera craniifer* Burm. (Dictyoptère). *C. R. Acad. Sci. Paris Ser. D 267: 1852-1855* **anat gut SGG SO dict**

DAOUD A USHERWOOD PNR 1975 Action of kainic acid on a glutamatergic synapse. *Comp. Biochem. Physiol. C 52: 51-53* **TS glu NM tox**

DAVENPORT AP 1985 The effect of stress on octopamine content of the insect haemolymph. *Pest. Sci. 16: 437-438* **TS blood oct rev**

DAVENPORT AP EVANS PD 1984 Stress-induced changes in the octopamine levels of insect haemolymph. *Insect Biochem. 14: 135-143* **TS blood oct**

DAVENPORT AP EVANS PD 1984 Changes in haemolymph octopamine levels associated with food deprivation in the locust *Schistocerca gregaria. Physiol. Entomol. 9: 269-274* **TS oct feed**

DAVENPORT AP EVANS PD 1986 Sex-related differences in the concentration of met-enkephalin-like immunoreactivity in the nervous system of an insect, *Schistocerca gregaria*, revealed by radioimmunoassay. *Brain Res. 238: 319-322* **TS pep enkeph**

DAVENPORT AP MORTON DB EVANS PD 1985 The action of formamidines on octopamine receptors in the locust. *Pest. Biochem. Physiol. 24: 45-52* **recept oct**

DAVID JC LAFON-CAZAL M 1979 Octopamine distribution in the *Locusta migratoria* nervous and non-nervous systems. *Comp. Biochem. Physiol. C 64: 161-164* **TS oct opt retroc**

DAVID OF GORELKIN VS SVIDERSKY VL 1977 [On the role of the receptors of the chordotonal organ of the forewings in the locust *Locusta migratoria].* (in Russian.) *Zh. evol. Biokhim. Fiziol. 13: 518-520* **flight CO thor SO**

DE LERMA B 1937 Osservazioni sul systema endocrino degli insetti (corpora allata e corpori faringei). *Arch. zool. ital. 24: 339-368* **retroc anat SGG CAll**

DE LOOF A HUYBRECHTS R VERHAERT P 1987 Vertebrate-peptide hormone like material in arthropods: identification, methods, and functions. *Bull. Serb. Acad. Sci. Arts 29: 25-35* **pep**

DE LOOF A SCHOOFS L 1989 Review: homologies between the amino acid sequences of some vertebrate-type peptide hormones and peptides isolated from invertebrate sources. *Comp. Biochem. Physiol. B 95: 495-468* **pep rev**

DE LOOF A SCHOOFS L HOLMAN GM HAYES TK NACHMAN RJ VANDEN BROECK J 1990 Isolation and identification of neuropeptides in *Locusta migratoria. In: McCaffery A Wilson I (eds.): Chromatography and isolation of insect hormones and pheromones. New York: Plenum Press. 205-211* **pep tech**

DE LOOF A VANDEN BROECK J SWEVERS L SCHOOFS L 1991 Vertebrate-type neuropeptides and steroids in *Locusta migratoria*: identification and metabolism. *In: Hagedorn HH Hildebrand JG Kidwell MG Law*

JH (eds.): Molecular Insect Science. New York: Plenum Press. 16-21 **pep hormX**

DE LOOF A VERHAERT P SCHOLS D THEUNIS W SCHOOFS L CLOTTENS F SWINNEN K DELANG I COOREMAN B 1986 Vertebrate-type brain and gastrointestinal peptides in *Periplaneta, Locusta* and *Crangon. Bull. Soc. zool. Fr. Belg. 111: 1-2* **pep dict**

DE LOOF A VERHAERT P SCHOOFS L HUYBRECHTS R FRANKENNE F HENNEN G VAUDRY H JEGOU S 1988 Vertebrate-type hormones in insects. *In: Sehnal F Zabza A Denlinger DL (eds.), Endocrinological frontiers in physiological insect ecology. Wroclaw: Technical University Press. 813-831* **pep hormX**

DEBAISIEUX P 1938 Organes scolopidiaux des pattes d'insectes. II. *Cellule 47: 79-202* **FCO SO propr**

DEBAUCHE H 1938 Etude cytologique et comparée de l'organe de Johnston des insectes. III. Les principaux organes sensoriels de l'antenne chez *Schistocerca gregaria* Forsk. *Proc. 5th. Int. Locust Conf. Brussels 1938: 417-433* **ant SO mech anat**

DEL CASTILLO J HOYLE G MACHNE X 1953 Neuromuscular transmission in a locust. *J. Physiol. (Lond.) 121: 539-547* **NM**

DELCOMYN F 1980 Neural basis of rhythmic behavior in animals. *Science 210: 492-498* **flight rhythm**

DELPHIN F 1963 The histology and possible functions of neurosecretory cells in the ventral ganglia of *Schistocerca gregaria* Forsk. *Nature 200: 913-915* **NSC thor abd ADH**

DELPHIN F 1965 The histology and possible functions of neurosecretory cells in the ventral ganglia of *Schistocerca gregaria* Forskål (Orthoptera: Acrididae). *Trans. R. Entomol. Soc. Lond. 117: 167-214* **NSC thor abd ADH**

DEY S 1988 Scanning electron microscopic detection of corneal anti-reflection coating in the grasshopper *Epacromia dorsalis* and its physiological significance. *Vision Res. 28: 975-977* **vis fineStruc**

DIEDEREN JHB MAAS HA PEL HJ SCHOONEVELD H JANSEN WF VULLINGS HGB 1987 Co-localization of the adipokinetic hormones I & II in the same glandular cells and in the same secretory granules of corpus cardiacum of *Locusta migratoria* and *Schistocerca gregaria*. An immuno-electron-microscopic study. *Cell Tiss. Res. 249: 379-389* **AKH pep fineStruc retroc CC**

DIEDEREN JHB VAN ETTEN EWM BIEGSTRAATEN AIM TERLOU M VULLINGS HGB JANSEN WF 1988 Flight-induced inhibition of the cerebral median peptidergic neurosecretory system in *Locusta migratoria. Gen. comp. Endocrinol. 71: 257* **pep flight brain NSC**

DIEDEREN JHB VULLINGS HGB JANSEN WF 1991 The functional significance of enhanced endocytosis in flight-stimulated adipokinetic cells in the corpus cardiacum of *Locusta migratoria. Proc. exper. appl. Entomol., N. E. V. Amsterdam 2: 92-93* **retroc CC AKH pep flight**

DIRCKSEN H BREIDBACH O 1991 Peripheral projection sites of crustacean cardioactive peptide (CCAP)-immunoreactive neurons of the central nervous system of the locust *Locusta migratoria* and the metamorphosing meal beetle *Tenebrio molitor. In: Elsner N Penzlin H (eds.), Synapse-transmission-modulation: Proc. 19th Göttingen Neurobiol. Conf. Stuttgart, New York: Thieme Verlag. Abstract 14* **pep CCAP anat**

DIRCKSEN H MÜLLER A KELLER R 1989 Immunocytochemical demonstration of CCAP, the novel cardioactive peptide, in the nervous system of the crab, *Carcinus maenas*, and the insect, *Locusta migratoria. Gen. comp. Endocrinol. 74: 241* **pep CCAP**

DIRCKSEN H MÜLLER A KELLER R 1989 Immunocytochemical demonstration of CCAP, the novel crustacean cardioactive peptide, in the nervous system of the locust, *Locusta migratoria. In: Elsner N Singer W (eds.), Dynamics and plasticity in neuronal systems: Proc. 17th. Göttingen Neurobiology Conference, Thieme Verlag. p. 59* **pep CCAP**

DIRCKSEN H MÜLLER A KELLER R 1991 Crustacean cardioactive peptide in the nervous system of the locust, *Locusta migratoria.* An immunocytochemical study on the ventral nerve cord and peripheral innervation. *Cell Tiss. Res. 263: 439-458* **pep CCAP anat**

DOE CQ BASTIANI MJ GOODMAN CS 1986 Guidance of neuronal growth cones in the grasshopper embryo. IV. Temporal delay experiments. *J. Neurosci. 6: 3552-3563* **devel**

DOE CQ GOODMAN CS 1985 Neurogenesis in grasshopper and fushi tarazu *Drosophila* embryos. *Cold Spring Harbour Symp. Quant. Biol. 50: 891-903* **devel**

DOE CQ GOODMAN CS 1985 Early events in insect neurogenesis: I. Development and segmental differences in the pattern of neuronal precursor cells. *Dev. Biol. 111: 193-205* **devel**

DOE CQ GOODMAN CS 1985 Early events in insect neurogenesis: II. The role of cell interactions and cell lineage in the determination of neuronal precursor cells. *Dev. Biol. 111: 206-219* **devel**

DOE CQ KUWADA J GOODMAN CS 1985 From epithelium to neuroblasts to neurones: the role of cell interactions and cell lineage during insect neurogenesis. *Phil. Trans. R. Soc. Lond. (Biol.) 312: 67-81* **devel**

DOERR-SCHOTT J JOLY L DUBOIS MP 1978 Sur l'existence dans la pars intercerebralis d'une insecte (*Locusta migratoria* R. et F.) de cellules neurosécrétrices fixant un antisérum antisomatostatine. *C. R. Acad. Sci. Paris Ser. D 286: 93-95* **NSC pep brain**

DOGRA GS EWEN AB 1970 Histology of the neurosecretory system and the retrocerebral endocrine glands of the adult migratory grasshopper, *Melanoplus sanguinipes* (Fab.) (Orthoptera: Acrididae). *J. Morphol. 130: 4512-466* **retroc anat CC CAll**

DONALDSON PL DUCE IR USHERWOOD PNR 1983 Calcium accumulation precedes the degenerative effects of L-glutamate on locust muscle fibres. *Brain Res. 261-265* **TS glu musc degen Ca2**

DOWNSON RJ USHERWOOD PNR 1973 The mode of action of phenylhydrazine hydrochloride on the locust neuromuscular system. *J. Insect Physiol. 19: 355-368* **TS glu NM tox**

DREHER A 1982 Optimierte Laborbedingungen für Langzeitflüge vor dem Windkanal bei der Wanderheuschrecke, *Locusta migratoria. In: Nachtigall W (ed.), BIONA-report 1 "Insekten Flug"; Akad. Wiss. Mainz. Stuttgart & New York: Gustav Fischer Verlag. 111-120* **flight tech**

DU LAC S BASTIANI MJ GOODMAN CS 1986 Guidance of neuronal growth cones in the grasshopper embryo. II. Recognition of a specific axonal pathway by the aCC neuron. *J. Neurosci. 6: 3532-3541* **devel**

DUBAS F 1989 Pharmacological studies on locust flight motoneurons. *In: Elsner N Singer W (eds.), Dynamics and plasticity in neuronal systems: Proc. 17th. Göttingen Neurobiology Conference. Stuttgart, New York: Thieme Verlag. Abstract 51* **flight MN thor TS GABA Glu ACh**

DUBAS F 1990 Action of glutamate on the neuropile arborizations of locust motoneurons. (Abstract). *Experientia 46: 75* **TS glu MN thor flight**

DUBAS F 1990 Amino acid neurotransmission in the neuropil arborizations of flight motoneurons in locusts. *In: Elsner N Roth G (eds.) Brain - Perception - Cognition: Proc. 18th Göttingen Neurobiol. Conf. Stuttgart, New York: Thieme Verlag. Abstract 493* **MN flight TS glu GABA**

DUBAS F 1990 Inhibitory effect of L-glutamate on the neuropile arborizations of flight motoneurones in locusts. *J. exp. Biol. 148: 501-508* **flight MN TS glu inhib**

DUBAS F 1990 Pharmacology of aminoacid receptors on locust flight motoneurons. *Rev. suisse Zool. 97: 806* **TS glu MN thor flight GABA recept**

DUBAS F 1990 Actions of glutamate, GABA and other putative aminoacid neurotransmitters on the neuropile arborizations of locust motoneurons. *Soc. Neurosci. Abstr. 16: 857* **TS glu MN thor flight GABA**

DUBAS F 1991 Actions of putative amino acid neurotransmitters on the neuropile arborizations of flight motoneurones. *J. exp. Biol. 155: 337-355* **TS glu MN thor flight GABA**

DUCE IR BUDD TC RICHARDSON PS 1991 The use of anti-idiotypic antibodies as a strategy for studying glutamate receptors. *Biochem. Soc. Trans. 19: 143-146* **musc anat glu tech recept**

DUCE IR DONALDSON PL USHERWOOD 1983 Investigations into the mechanism of excitant aminoacid cytotoxicity using a well-characterized glutamatergic system. *Brain Res. 263: 77-87* **TS glu tox**

DUCE IR SCOTT RH 1984 GABA-sensitivity in the distal bundles of the locust extensor tibiae muscle. *J. Physiol. (Lond.) 343: 31-32P* **musc TS GABA**

DUCE JA MILLER BA USHERWOOD PNR 1988 Patch-clamp studies of cultured muscle from locust embryos. *In: Beadle DJ Lees G Kater SB (eds.), Cell culture approaches to invertebrate neuroscience. London & New York: Academic Press. 221-233* **TS glu musc cellCult devel**

DUCE JA USHERWOOD PNR 1986 Primary cultures of muscle from embryonic locusts (*Locusta migratoria & Schistocerca gregaria*): developmental, electrophysiological and patch clamp studies. *J. exp. Biol. 123: 307-323* **cellCult musc TS devel**

DUDEL J FRANKE C HATT H RAMSEY RL USHERWOOD PNR 1988 Rapid activation and desensitization by glutamate of excitatory, cation-selective channels in locust muscle. *Neurosci. Lett. 88: 33-38* **channel musc glu**

DUDEL J FRANKE C HATT H RAMSEY RL USHERWOOD PNR 1990 Glutaminergic channels in locust muscle show a wide time range of desensitization and resensitization characteristics. *Neurosci. Lett. 114: 207-212* **channel musc glu desens**

DUDEL J FRANKE C HATT H USHERWOOD PNR 1989 Chloride channels gated by extrajunctional glutamate receptors (H-receptors) on locust leg muscle. *Brain Res. 481: 215-220* **TS glu musc channels recept**

DUERR JS CONDIC ML 1990 Pioneer neuron guidance cues in grasshopper limb buds. *Soc. Neurosci. Abstr. 16: 810* **devel leg**

DUGARD JJ 1967 Directional change in flying locusts. *J. Insect Physiol. 13: 1055-1063* **flight behav leg aero OM**

DUNBAR S HUDDART H 1982 Calcium movements in insect visceral muscle. *Comp. Biochem. Physiol. A 71: 425-437* **gut musc Ca2 glu proct pep saline**

DUNBAR SJ PIEK T 1982 The action of *Philanthus triangulum* venom on electro-mechanical responses of a locust (*Locusta migratoria*) hindgut muscle. *Proc. Ned. Fed. 23: 111* **musc gut TS glu**

DUNBAR SJ PIEK T 1983 The action of iontophoretically applied L-glutamate on an insect (*Locusta migratoria migratorioides*) visceral muscle. *Archs. Insect Biochem. Physiol. 1: 93-104* **musc gut TS glu**

EBNER I BÄSSLER U 1978 Zur Regelung der Stellung des Femur-Tibia-Gelenkes im Mesothorax der Wanderheuschrecke *Schistocerca gregaria* (Forskål). *Biol. Cybern. 29: 83-96* **propr leg**

ECKERT M VARANKA I BENEDECZKY I 1989 Proctolinergic innervation of the locust oviduct. *Zool. Jb. (Allg. Zool. Physiol.) 93: 471-479* **genit pep proct innerv**

EGGERS A PREISS R GEWECKE M 1989 The characteristic curve of optomotor yaw-torque response in tethered flying locusts. *In: Elsner N Singer W (eds.), Dynamics and plasticity in neuronal systems: Proc. 17th. Göttingen Neurobiology Conference. Stuttgart, New York: Thieme Verlag. Abstract 111* **flight OM vis oc**

EGGERS A PREISS R GEWECKE M 1991 The optomotor yaw response of the desert locust, *Schistocerca gregaria. Physiol. Entomol. 16: 411-418* **flight behav OM**

EGGERS F 1928 Die stiftführenden Sinnesorgane. Morphologie und Physiologie der chordotonalen und tympanalen Sinnesapparate der Insekten. *Zoologische Bausteine 2: 1-354* **SO tymp anat CO**

EICHENDORF A KALMRING K 1980 Projections of auditory ventral cord neurons in the supraoesophageal ganglion of *Locusta migratoria. Zoomorphol. 94: 133-292* **aud IN brain anat**

ELDEFRAWI AT ELDEFRAWI ME KONNO K MANSOUR NA NAKANISHI K OLTZ E USHERWOOD PNR 1988 Structure and synthesis of a potent glutamate receptor antagonist in wasp venom. *Proc. nat. Acad. Sci. USA 85: 4910-4913* **TS glu musc tox recept**

ELIAS MS EVANS PD 1983 Histamine in the insect nervous system: distribution, synthesis and metabolism. *J. Neurochem. 41: 708-721* **TS devel hist metab**

ELIAS MS EVANS PD 1984 Autoradiographic localization of 3H-histamine accumulation by the visual system of the locust. *Cell Tiss. Res. 238: 105-112* **TS ret opt hist anat**

ELIAS MS LUMMIS SCR EVANS PD 1984 {3H}-mepyramine binding sites in the optic lobe of the locust: autoradiographic and pharmacological studies. *Brain Res. 294, 359-362* **TS ret opt hist tox**

ELLIOT RH GILLOTT C 1978 The neuroendocrine control of protein metabolism in the migratory grasshopper *Melanoplus sanguinipes. J. Insect Physiol. 24: 119-126* **pep metab**

ELLIS PE HOYLE G 1954 A physiological interpretation of the marching of hoppers of the african migratory locust (*Locusta migratoria migratorioides* R. and F.). *J. exp. Biol. 32: 271-279* **blood behav**

ELOFSSON R KLEMM N 1972 Monoamine-containing neurons in the optic ganglia of crustaceans and insects. *Z. Zellforsch. mikr. Anat. 133: 475-499* **TS opt anat 5HT norad**

ELSNER N 1967 Muskelaktivität und Lauterzeugung bei einer Feldheuschrecke. *Verh. dt. zool. Ges. 1967: 592-601* **strid musc behav**

ELSNER N 1968 Die neuromuskulären Grundlagen des Werbeverhaltens der Roten Keulenheuschrecke *Gomphocerippus rufus* (L.) *Z. vergl. Physiol. 60: 308-350* **strid**

ELSNER N 1970 Kommandofasern im Zentralnervensystem der Heuschrecke *Gastrimargus africanus* (Oedipodinae). *Verh. dt. zool. Ges. 1969: 465-471* **flight walk DN**

ELSNER N 1970 Die Registrierung der Stridulationsbewegungen bei der Feldheuschrecke *Chorthippus mollis* mit Hilfe von Hallgeneratoren. *Z. vergl. Physiol. 68: 417-428* **strid leg tech**

ELSNER N 1973 The central nervous control of courtship behaviour in the grasshoppper *Gomphocerippus rufus* (L.) (Orthoptera: Acrididae.) *In: Salánki J (ed..), Invertebrate Neurobiology: Mechanisms of rhythm regulation. Budapest: Kiadó. 1-20* **strid SO**

ELSNER N 1974 Neuroethology of sound production in Gomphocerine grasshoppers (Orthoptera: Acrididae). I. Song patterns and stridulatory movement. *J. comp. Physiol. 88: 67-102* **strid SO aud**

ELSNER N 1975 Neuroethology of sound production in Gomphocerine grasshoppers (Orthoptera: Acrididae). II. Neuromuscular activity underlying stridulation. *J. comp. Physiol. A 97: 291-322* **strid**

ELSNER N 1981 Developmental aspects of insect neuroethology. *In: Immelmann K Barlow GB Petrinovich L Main N (eds.), Behavioural de-*

velopment. Cambridge: Cambridge University Press. 474-490 **strid devel flight rev**

ELSNER N 1983 A neuroethological approach to the phylogeny of leg stridulation in gomphocerine grasshoppers. *In: Huber F Markl H (eds.), Neuroethology and behavioural physiology. Berlin, Heidelberg, New York: Springer Verlag. 53-86* **strid flight rhythm**

ELSNER N GRAMER R 1991 Motor patterns of flight and leg stridulation in acridid grasshoppers. *In: Elsner N Penzlin H (eds.), Synapse-transmission-modulation: Proc. 19th Göttingen Neurobiol. Conf. Stuttgart, New York: Thieme Verlag. Abstract 143* **behav strid flight rhythm**

ELSNER N HIRTH C 1981 Short- and long-term control of motor coordination in a stridulating grasshopper. *Naturwiss. 65: 160* **strid**

ELSNER N HUBER F 1969 Die Organisation des Werbegesanges der Heuschrecke *Gomphocerippus rufus* in Abhängigkeit von zentralen und peripheren Bedingungen. *Z. vergl. Physiol. 65: 389-423* **strid**

ELSNER N HUBER F 1973 Neurale Grundlagen artspezifischer Kommunikation bei Orthopteren. *Fortschr. Zool. 22: 1-48* **strid behav rev**

ELSNER N POPOV AV 1978 Neuroethology of acoustic communication. *Adv. Insect Physiol. 13: 229-335* **strid aud rev**

ELSNER N VON HELVERSEN O 1976 Altersbedingte Abwandlungen der Stridulationsmuster bei Feldheuschrecken. *Verh. dt. zool. Ges. 1976: 280* **strid**

ELSON RC 1987 Flight motor neurone reflexes driven by strain-sensitive wing mechanoreceptors in the locust. *J. comp. Physiol. A 161: 747-760* **flight propr MN thor SO mech**

ELSON RC 1987 Interneuronal processing of inputs from the campaniform sensilla of the locust hindwing. *J. comp. Physiol. A 161: 761-776* **propr flight IN thor MN SO mech**

ELSON RC 1987 Integration of wing proprioceptive and descending exteroceptive sensory inputs by thoracic interneurones of the locust. *J. exp. Biol. 128: 193-218* **flight SO DN thor IN propr**

ELSON RC PFLÜGER H-J 1986 The activity of a steering muscle in flying locusts. *J. exp. Biol. 120: 421-442* **flight musc**

EMERY DJ BELL KA CHAPCO W STEEVES JD 1984 Characterization of a reduced-eye mutant of the grasshopper *Melanoplus sanguinipes*. *J. Embryol. exp. Morphol. 83: 189-211* **vis genet DN oc anat**

EMSON PC BURROWS M FONNUM F 1974 Levels of glutamate decarboxylase, choline acetyl transferase and acetylcholine esterase in identified motor neurones of the locust. *J. Neurobiol. 5: 33-42* **TS enz ACh Glu**

ERIKSSON ES 1980 Movement parallax and distance perception in the grasshopper *Phaulacridium vittatum* (Sjöstedt). *J. exp. Biol. 86: 337-340* **vis**

ERNST KD BOEKH J BOEKH V 1977 A neuroanatomical study on the organization of the central antennal pathways in insects. II. Deutocerebral connections in *Locusta migratoria* and *Periplaneta americana. Cell Tiss. Res. 176: 285-308* **ant brain anat dict**

EVANS ML USHERWOOD PNR 1985 The efffect of lectins on desensitisation of locust muscle glutamate receptors. *Brain Res. 358: 34-39* **TS glu musc recept desens tox**

EVANS PD 1978 Octopamine; from metabolic mistake to modulator. *Trends in Neurosci. 1: 154-157* **TS mod oct rev**

EVANS PD 1980 Biogenic amines in the insect nervous system. *Adv. Insect Physiol. 15: 317-473* **TS rev**

EVANS PD 1980 The biochemistry of insect neurohormones, with special reference to octopamine, a neuromodulator. *In: Sherwood M (ed.): Insect neurobiology and pesticide action (Neurotox 79). London: Society of Chemical Industry Press. 313-320* **TS mod oct**

EVANS PD 1980 Octopamine receptors in insects. *In: Sattelle DB Hall LM Hildebrand JG (eds.): Receptors for neurotransmitters, hormones and pheromones in insects. Amsterdam: Elsevier/North-Holland Biomedical Press* **TS mod oct recept**

EVANS PD 1981 Octopaminergic modulation of neuromuscular transmission in the locust. *In: Neurotransmitters in invertebrates. Hungarian Academy of Sciences Academic Press. 283-303* **TS mod oct NM**

EVANS PD 1981 Multiple receptor types for octopamine in the locust. *J. Physiol. (Lond.) 318: 99-122* **TS oct recept**

EVANS PD 1982 Properties of modulatory octopamine receptors in the locust. *In: Evered D O'Connor M Whelan J (eds.), Neuropharmacology of insects. (Ciba Foundation Symp. 88). London: Pitman. 48-69* **musc TS oct mod recept**

EVANS PD 1984 Studies on the mode of action of octopamine, 5-hydroxytryptamine and proctolin on a myogenic rhythm in the locust. *J. exp. Biol. 110: 231-251* **TS oct 5HT pep rhythm leg mod**

EVANS PD 1984 A modulatory octopaminergic neurone increases cyclic nucleotide levels in locust skeletal muscle. *J. Physiol. (Lond.) 348: 307-324* **TS oct DUM musc cAMP mod**

EVANS PD 1984 The role of cyclic nucleotides and calcium in the mediation of the modulatory effects of octopamine on locust skeletal muscle. *J. Physiol. (Lond.) 348: 325-340* **cAMP TS oct musc mod**

EVANS PD 1985 Biogenic amines and second messenger systems in insects. *In: Keyserlingk HC Jäger A von Szczepanski C (eds), Approaches to new leads for insecticides. Berlin, Heidelberg, New York: Springer Verlag. 117-131* **TS cAMP**

EVANS PD 1985 Octopamine. *In: Kerkut GA Gilbert LI (eds.): Comprehensive insect physiology biochemistry and pharmacology. Vol. II. London: Pergamon Press. 499-530* **oct rev**

EVANS PD 1985 Regional differences in responsiveness to octopamine within a locust skeletal muscle. *J. Physiol. (Lond.) 366: 331-341* **TS oct mod musc**

EVANS PD 1986 Biogenic amine receptors and their mode of action in insects. *In: Borkovec AB Gelman DB (eds.), Insect neurochemistry and neurophysiology. Clifton, N.J.: Humana Press. 117-141* **TS cAMP recept 5HT oct**

EVANS PD 1987 Phenyliminoimidazolidine derivatives activate both octopamine 1 and octopamine 2 receptor subtypes in locust skeletal muscle. *J. exp. Biol. 129: 239-250* **musc TS oct recept tox**

EVANS PD 1988 Octopamine receptors in locust skeletal muscle. *In: Boulton AA Juorio AV Downer RGH (eds.): Trace amines: Comparative, neurobiological and clinical studies. Clifton, New Jersey: Humana Press. 17-31* **TS oct musc flight mod recept**

EVANS PD COURNIL I 1990 Co-localization of FLRF- and vasopressin-like immunoreactivity in a single pair of sexually dimorphic neurones in the nervous system of the locust. *J. comp. Neurol. 292: 331-348* **pep anat IN FMRF vasop suboes dimorph**

EVANS PD DAVENPORT AP 1986 The action of pesticides on biogenic amine receptors and second messenger systems in insects. *In: Ford MG Lung GG Reay RC Usherwood PNR (eds.), Neuropharmacology and pesticide action. Chichester: Ellis Horwood. 315-341* **TS cAMP oct tox**

EVANS PD DAVENPORT AP ELIAS MS MORTON DB TRIMMER BA 1985 Assays for biogenic amines in insect nervous tissue. *In: Breer H Miller TA (eds.): Neurochemical techniques in insect research. Berlin, Heidelberg, New York: Springer Verlag. 25-46* **TS tech oct hist 5HT**

EVANS PD GEE JD 1980 Action of formamidine pesticides on octopamine receptors. *Nature 287: 60-62* **TS oct recept tox**

EVANS PD MYERS CM 1986 Peptidergic and aminergic modulation of insect skeletal muscle. *J. exp. Biol. 124: 143-176* **TS mod pep proct oct musc**

EVANS PD MYERS CM 1986 The modulatory actions of FMRFamide and related peptides on locust skeletal muscle. *J. exp. Biol. 126: 403-422* **TS pep musc FMRF mod**

EVANS PD O'SHEA M 1977 The identification of an octopaminergic neurone which modulates neuromuscular transmission in the locust. *Nature 270: 275-279* **TS oct NM mod**

EVANS PD O'SHEA M 1978 The identification of an octopaminergic neurone and the modulation of a myogenic rhythm in the locust. *J. exp. Biol. 73: 235-260* **TS oct musc mod DUM rhythm**

EVANS PD ROBB S COURNIL I 1990 Characterization of FMRFamide-like peptides in the locust, *Schistocerca gregaria. In: Borkovec AB Masler EP (eds.), Insect neurochemistry and neurophysiology. Clifton, N.J.: Humana Press Inc. 397-400* **pep FMRF**

EVANS PD ROBB S CUTHBERT BA 1988 Modulation of neuromuscular transmission in the locust by FMRFamide-like peptides. *In: Herting G Spatz H-C (eds.), Modulation of synaptic transmission and plasticity in nervous tissues. Berlin, Heidelberg, New York: Springer Verlag & NATO scientific affairs division. 305-318* **pep FMRF NM mod**

EVANS PD ROBB S CUTHBERT BA 1989 Insect neuropeptides - identification, establishment of functional roles and novel target sites for pesticides. *Pest. Sci. 25: 71-83* **pep TOX**

EVANS PD SIEGLER MVS 1982 Octopamine mediated relaxation of maintained and catch tension in locust skeletal muscle. *J. Physiol. (Lond.) 324: 93-112* **TS oct musc MOD**

EVANS PD SWALES LS WHIM MD 1988 Second messenger systems in insects: an introduction. *In: Lunt GG (ed.): Neurotox '88: Molecular basis of drug and pesticide action. Amsterdam: Elsevier Science Publishers BV. 225-234* **cAMP lip**

EVANS PD THONOOR CM MIDGLEY JM 1988 Activities of octopamine and synephrine stereoisomers on octopaminergic receptor subtypes in locust skeletal muscle. *J. Pharm. Pharmacol. 40: 855-861* **TS oct musc recept**

EWER DW 1953 The anatomy of the nervous system of the tree locust *Acanthacris ruficornis* (Fab.) I. The adult metathorax. *Ann. Natal Mus. 12: 367-381* **anat musc thor**

EWER DW 1954 On the nymphal musculature of the pterothorax of certain Acrididae (Orthoptera). *Ann. Natal Mus. 13: 79-89* **anat musc thor pyrg**

EWER DW 1954 The anatomy of the nervous system of the tree locust *Acanthacris ruficornis* (Fab.). II. The adult mesothorax. *J. entomol. Soc. S. Afr. 17: 27-37* **anat musc thor**

EWER DW 1954 The anatomy of the nervous system of the tree locust *Acanthacris ruficornis* (Fab.). III. The innervation of the nymphal muscles of the pterothorax and the first abdominal segment. *J. entomol. Soc. S. Afr. 17: 232-236* **anat musc thor devel**

EWER DW 1954 A note on the comparative anatomy of the pterothorax of macropterous and brachypterous forms of the grasshopper, *Zonocerus elegans* Thunb. *J. entomol. Soc. S. Afr. 17: 237-240* **anat musc thor pyrg**

EWER DW 1954 A note on the comparative anatomy of the pterothorax of macropterous and brachypterous forms of the grasshopper *Zonocerus elegans* Thunb. *J. entomol. Soc. S. Afr. 17: 237-240* **anat musc thor pyrg**

EWER DW 1955 Notes on acridid anatomy. I. The prothoracic musculature of certain acridids. *J. entomol. Soc. S. Afr. 18: 42-47* **anat musc thor head pyrg**

EWER DW 1957 The anatomy of the nervous system of the tree locust *Acanthacris ruficornis* (Fab.). IV. The prothorax. *J. entomol. Soc. S. Afr. 20: 195-204* **anat musc thor**

EWER DW 1957 The anatomy of the nervous system of the tree locust *Acanthacris ruficornis* (Fab.). V. The homologies of the thoracic musculature. *J. entomol. Soc. S. Afr. 20: 204-216* **anat musc thor**

EWER DW 1957 Notes on acridid anatomy. II. A sexual dimorphism in the thorax of certain acridids. *J. entomol. Soc. S. Afr. 20:229-231* **anat musc thor pyrg dimorph**

EWER DW 1957 Notes on acridid anatomy. III. Twitch and tonic muscles in the thorax of *Locusta. J. entomol. Soc. S. Afr. 20: 231-233* **anat musc thor**

EWER DW 1957 Notes on acridid anatomy. IV. The anterior abdominal musculature of certain acridids. *J. entomol. Soc. S. Afr. 20: 260-79* **anat musc abd pyrg**

EWER DW 1958 Notes on acridid anatomy. V. The pterothoracic musculature of *Lentula callani* Dirsch. *J. entomol. Soc. S. Afr. 21: 132-138* **anat musc thor lent**

EWER DW 1964 Notes on acridid anatomy. VI. On the pterothoracic musculature of *Bullacris* Roberts and *Pneumora* Stål (Orthoptera: Pneumoridae). *J. entomol. Soc. S. Afr. 26: 411-424* **anat musc thor pneu**

EWER DW RIPLEY SH 1953 On certain properties of the flight muscles of the Orthoptera. *J. exp. Biol. 30: 170-177* **flight musc**

FACCIPONTE G LANGE AB 1991 The rhythmic motor pattern controlling oviduct contractions in *Locusta migratoria*: descending neural control and tracing of efferent targets. *Soc. Neurosci. Abstr. 17: 1056* **abd genit rhythm innerv musc**

FAHIM MA USHERWOOD PNR 1983 Effects of c-AMP, caffeine, theophylline and vinblastine on spontaneous transmitter release at locust nerve-muscle junctions. *J. Neurobiol. 14: 391-397* **TS glu NM cAMP**

FARROW RA 1990 Flight and migration in acridoids. *In: Chapman RF Joern A (eds.), Biology of grasshoppers. New York: J. Wiley & Sons Inc. 227-314* **flight behav**

FEKETE E HALASY K GABRIEL R BENEDECZKY I 1988 Some histochemical aspects of invertebrate enteric nervous system. *In: Salánki J S-Rózsa K (eds.), Neurobiology of invertebrates. (Symp. Biol. Hung. 36). Budapest. 185-197* **gut**

FERBER M 1989 Are all DUM-neurones octopaminergic? Evidence for peptidergic DUM-neurones in locust abdominal ganglia. *In: Erber J Menzel Pflüger H-J Todt D (eds.), Neural mechanisms of behavior: Proc. 2nd Internat. Congr. Neuroethology. Stuttgart & New York: Thieme Verlag. Abstract 249* **DUM TS oct pep abd**

FERBER M PFLÜGER H-J 1988 Abdominal DUM-neurones: target muscles and modulatory actions. *In: Elsner N Barth FG (eds.), Sense organs: interfaces between environment and behaviour: Proc.16th. Göttingen Neurobiol. Conf. Stuttgart, New York: Thieme Verlag. Abstract 103* **DUM abd heart MN anat musc oct TS**

FERBER M PFLÜGER H-J 1989 A population of DUM-like neurons within the abdominal ganglia of the locust. *In: Elsner N Singer W (eds.), Dynamics and plasticity in neuronal systems: Proc. 17th. Göttingen Neurobiology Conference. Stuttgart, New York: Thieme Verlag. Abstract 57* **DUM abd heart anat**

FERBER M PFLÜGER H-J 1990 Bilaterally projecting neurones in pregenital abdominal ganglia of the locust: anatomy and peripheral targets. *J.*

comp. Neurol. 302: 447-460 **DUM abd heart MN anat musc oct TS innerv**

FERNANDEZ-MORAN H 1958 Fine structure of the light receptors in the compound eyes of insects. *Exp. Cell Res. (Suppl.) 5: 586-644* **ret fineStruc**

FIELD LH 1991 Mechanism for range fractionation in chordotonal organs of *Locusta migratoria* (L.) and *Valanga* spp. (Orthoptera: Acrididae). *Int. J. Insect Morphol. Embryol. 20: 25-39* **FCO SO propr leg**

FIELD LH BURROWS M 1982 Reflex effects of the femoral chordotonal organ upon leg motor neurones of the locust. *J. exp. Biol. 101: 265-286* **FCO leg MN SO propr**

FIELD LH PFLÜGER H-J 1989 The femoral chordotonal organ: a bifunctional orthopteran (*Locusta migratoria*) sense organ? *Comp. Biochem. Physiol. A 93: 729-743* **FCO propr anat thor vibr SO leg**

FIELD LH RIND FC 1981 A single insect chordotonal organ mediates inter- and intra-segmental leg reflexes. *Comp. Biochem. Physiol. A 68: 99-102* **FCO propr SO leg**

FISHER EJ PATTERSON JA 1988 A simple, inexpensive method for plotting receptive fields of locust photoreceptors using a video monitor and a BBC microcomputer. *J. Physiol. (Lond.) 399: 3P* **tech ret**

FONSECA MI LUNT GG AGUILAR JS 1991 Inhibition of muscarinic cholinergic receptors by disulfide reducing agents and arsenicals. Differential effect on locust and rat. *Biochem. Pharmacol. 41: 735-742* **brain ACh recept tox**

FORD N 1923 A comparative study of the abdominal musculature of orthopteroid insects. *Trans. R. Can. Inst. 14: 207-319* **abd musc anat**

FORD WCL CANDY DJ 1972 The regulation of glycolysis in perfused locust flight muscle. *Biochem. J. 130: 1101-1112* **metab musc flight**

FORMAN RR 1984 Leg position learning by an Insect. I. A heat avoidance learning paradigm. *J. Neurobiol. 15: 127-140* **leg learn behav**

FORMAN RR ZILL SN 1984 Leg position learning by an Insect. II. Motor strategies underlying learned leg extension. *J. Neurobiol. 15: 221-237* **leg learn behav**

FOSTER JA ROBERTSON RM 1991 Temperature dependency of wingbeat frequency in intact and deafferented locusts. *J. exp. Biol.* 162: 295-31 **flight temp**

FOURNIER B 1990 Neuroparsins induce phosphoinositide breakdown in the migratory locust rectal cells. *Comp. Biochem. Physiol. B 95: 57-64* **pep neuropars insP gut**

FOURNIER B 1990 Functional cross-reactivities and transduction systems at the rectal level by antidiuretic peptides from cockroach and locust corpora cardiaca. *J. Insect Physiol. 36: 537-544* **pep ADH gut dict**

FOURNIER B DUBAR M 1989 Relationship between neuroparsin-induced rectal fluid reabsorption and cyclic nucleotides in the migratory locust. *Comp. Biochem. Physiol. A 94: 249-255* **neuropars cAMP gut**

FOURNIER B GIRARDIE J 1988 A new function for the locust neuroparsins: stimulation of water reabsorption. *J. Insect Physiol. 34: 309-313* **pep neuropars ADH**

FOURNIER B HERAULT JP PROUX J 1987 Antidiuretic factor from the nervous corpora cardiaca of the migratory locust: improvement of an existing in vitro bioassay. *Gen. comp. Endocrinol. 68: 49-56* **pep ADH retroc CC**

FRAENKEL G 1932 Untersuchungen über die Koordination von Reflexen und automatisch-nervösen Rhythmen bei Insekten. I. Die Flugreflexe von Insekten und ihre Koordination. *Z. vergl. Physiol. 16: 371-393* **rhythm flight**

FRAENKEL G 1932 Untersuchungen über die Koordination von Reflexen und automatisch-nervösen Rhythmen bei Insekten. IV. Über die nervösen Zentren der Atmung und die Koordination ihrer Tätigkeit. *Z. vergl. Physiol. 16: 418-443* **rhythm resp spir**

FRAENKEL G 1932 Untersuchungen über die Koordination von Reflexen und automatisch-nervösen Rhythmen bei Insekten. III. Das Problem des gerichteten Atemstromes in den Tracheen der Insekten. *Z. vergl. Physiol. 16: 418-443* **rhythm resp spir trach**

FRANKLIN R 1975 Swimming coordination in *Melanoplus differentialis* (Abstr.). *J. Kansas entomol. Soc. 48: 502* **jump swim**

FRANKLIN R JANDER R ELE K 1977 The coordination, mechanics, and evolution of swimming by a grasshopper, *Melanoplus differentialis* (Orthoptera, Acrididae). *J. Kansas entomol. Soc. 50: 189-199* **swim**

FRASER SP DJAMGOZ MBA USHERWOOD PNR O'BRIEN J DARLISON MG BARNARD EA 1990 Aminoacid receptors from insect muscle - electrophysiological characterisation in *Xenopus* oocytes following expression by injection of messenger RNA. *Mol. Brain Res. 8: 331-341* **recept musc**

FREEMAN MA 1966 The effects of drugs on the alimentary canal of the african migratory locust, *Locusta migratoria. Comp. Biochem. Physiol. 17: 755-764* **gut 5HT**

FRENCH AS KUSTER JE 1985 Nonlinearities in locust photoreceptors during transduction of small numbers of photons. *J. comp. Physiol. A 156: 645-652* **ret**

FRIEDEL T LOUGHTON BG 1980 The synthesis of neurosecretory protein during the fifth instar of *Locusta migratoria migratorioides. Gen. comp. Endocrinol. 41: 261-267* **brain retroc NSC pep**

FRIEDEL T LOUGHTON BG ANDREW RD 1980 A neurosecretory protein from *Locusta migratoria. Gen. comp. Endocrinol. 41: 487-498* **brain retroc NSC pep**

FRIEDEL T ORCHARD I LOUGHTON BG 1981 Release of neurosecretory protein from insect neurohaemal tissue following electrical stimulation. *Brain Res. 208: 451-455* **NSC pep brainStim retroc**

FRIEDRICH H 1929 Vergleichende Untersuchungen über die tibialen Scolopalorgane einiger Orthopteren. *Z. wiss. Zool. 134: 84-148* **SO leg mech CO**

FULTON BP USHERWOOD PNR 1977 Presynaptic acetylcholine action at the neuromuscular junction. *Neuropharmacol. 16: 877-880* **TS ACh NM mod**

FUZEAU-BRAESCH S 1981 Influence de la naloxone sur la perception de la presence du congénère par le criquet migrateur *Locusta migratoria. Agressologie 22: 195-197* **pep behav**

FUZEAU-BRAESCH S 1983 Polymorphisme phasaire et rôle du gaz carbonique chez *Locusta migratoria*, discussion de deux méthodes d'analyse: morphometrie et endocrinologie. *Rev. Can. Biol. Exp. 42: 223-227* **phase**

FUZEAU-BRAESCH S 1984 L'Octopamine: Erreur metabolique ou vestige evolutif? *Agressologie 25: 1-2* **oct**

FUZEAU-BRAESCH S 1986 Circadian changes of brain octopamine in relation with locomotor activity in the locust *Locusta migratoria cinerascens (gregaria)*. (Abstract) *Bull. Soc. zool. Fr. 111: 37* **rhythm oct behav**

FUZEAU-BRAESCH S COULON JF DAVID JC 1979 Octopamine levels during the moult cycle and adult development in the migratory locust, *Locusta migratoria. Experientia 35: 1349-1350* **TS oct moult**

FUZEAU-BRAESCH S NICOLAS G 1980 Effect of carbon dioxide on subsocial insects. *Comp. Biochem. Physiol. A 68: 289-297* **rev phase**

FUZEAU-BRAESCH S PAPIN C 1983 Dosage de l'octopamine par chromatographie liquide a haute performance (HPLC) et detection electrochimique (ED): essais sur le cerveau du criquet *Locusta migratoria cinerascens. Agressologie 24: 377-379* **brain oct**

GABRIEL JH 1984 The effect of animal design on jumping performance. *J. Zool. (Lond.) 204: 533-539* **jump**

GABRIEL JH 1985 The development of the locust jumping mechanism. I. Allometric growth and its effect on jumping performance. *J. exp. Biol. 118: 313-326* **jump**

GABRIEL JH 1985 The development of the locust jumping mechanism. II. Energy storage and muscle mechanics. *J. exp. Biol. 118: 327-340* **jump**

GABRIEL R HALASY K CSOKNYA M 1988 Visualization of neurons by NADH-diaphorase staining in the myenteric plexus of some invertebrate and vertebrate species. *Z. mikr.-anat. Forsch. (Leipzig) 102: 769-784* **gut anat enz**

GÄDE G 1979 Studies on the influence of synthetic adipokinetic hormone and some analogs on cyclic AMP levels in different arthropod systems. *Gen. comp. Endocrinol. 37: 122-130* **AKH cAMP**

GÄDE G 1981 Activation of fat body glycogen phosphorylase in *Locusta migratoria* by corpus cardiacum extract and synthetic adipokinetic hormone. *J. Insect Physiol. 27: 155-161* **pep retroc AKH enz metab**

GÄDE G 1987 Isolation and amino acid composition of two new neuropeptides from the corpus cardiacum of the eastern lubber grasshopper, *Romalea microptera*, belonging to the AKH/RPCH peptide family. *Zool. Jb. (Allg. Zool. Physiol.) 91: 307-303* **pep retroc AKH rom**

GÄDE G 1990 The adipokinetic hormone/red pigment concentrating hormone peptide family: structure, interrelationships and functions. *J. Insect Physiol. 36: 1-12* **pep AKH**

GÄDE G 1990 Structure-function studies on hypertrehalosaemic and adipokinetic hormones: activity of naturally occurring analogues and some N- and C- terminal modified analogues. *Physiol. Entomol. 15: 299-316* **pep AKH**

GÄDE G BEENAKKERS AMTH 1977 Adipokinetic hormone-induced lipid mobilization and cyclic AMP accumulation in the fat body of *Locusta migratoria* during development. *Gen. comp. Endocrinol. 32: 481-487* **AKH cAMP devel**

GÄDE G GOLDSWORTHY GJ KEGEL G KELLER R 1984 Single step purification of locust adipokinetic hormones I and II by reversed-phase high performance liquid chromatography, and amino-acid composition of the hormone II. *Hoppe-Seyler's Z. Physiol. Chem. 365: 393-398* **pep AKH tech**

GÄDE G GOLDSWORTHY GJ SCHAFFER MXH COOK JC RINEHART KL 1986 Sequence analyses of adipokinetic hormones II from corpora cardiaca of *Schistocerca nitens*, *Schistocerca gregaria*, and *Locusta migratoria* by fast atom bombardment mass spectrometry. *Biochem. Biophys. Res. Comm. 134: 723-730* **pep AKH tech**

GÄDE G HILBICH C BEYREUTHER K RINEHART KL 1988 Sequence analyses of two neuropeptides of the AKH/RPCH-family from the lubber grasshopper *Romalea microptera*. *Peptides (Fayetteville) 9: 681-688* **pep retroc AKH rom**

GÄDE G HOLWERDA DA 1976 Involvement of adenosine 3', 5'-cyclic monophosphate for lipid mobilization in *Locusta migratoria*. *Insect Biochem. 6: 535-540* **AKH cAMP flight**

GÄDE G SPRING JH 1989 Activation of fat body glycogen phosphorylase in the eastern lubber grasshopper (*Romalea microptera*) by the endogenous neuropeptides Ro I & Ro II. *J. exp. Zool. 250: 140-149* **retroc pep NSC brain AKH rom enz metab**

GAFFAL KP TICHY H THEISS J SEELINGER G 1975 Structural polarities in mechanosensitive sensilla and their influence on stimulus transmission (Arthropoda). *Zoomorphol. 82: 79-103* **windh SO mech fineStruc**

GALLOWAY JS RUNION HI USHERWOOD PNR 1966 An electrophysiological approach to the study of insect behaviour. *J. Physiol. (Lond.) 187: 9-10P* **walk**

GAWANDE BR 1969 Effects of implanted endocrine organs on the neuro-endocrine system of the host: experiments on nymphs of *Schistocerca gregaria* . *Acta entomol. bohem. 66: 266-274* **NSC brain retroc CC CAll**

GAWANDE BR 1969 Neurosecretion in the Vth instar nymphs of *Schistocerca gregaria* Forsk. *Acta entomol. bohem. 66: 70-80* **NSC brain retroc CC CAll**

GEFFARD M VIEILLEMARINGE J HEINRICH-ROCK AM DURIS P 1985 Anti-acetylcholine antibodies and first immunocytochemical application in insect brain. *Neurosci. Lett. 57: 1-6* **TS ACh brain anat**

GEIST R 1928 The heat-sensitive areas of certain grasshoppers. *Ann. entomol. Soc. Am. 21: 614-618* **SO therm**

GELDIAY S 1967 Hormonal control of adult reproductive diapause in the egyptian grasshopper *Anacridium aegyptium* L. *J. Endocrinol. 37: 63-71* **hormX diap**

GELDIAY S 1970 Photoperiodic control of neurosecretory cells in the brain of the egyptian grasshopper *Anacridum aegyptium* L. *Gen. comp. Endocrinol. 14: 35-42* **NSC hormX brain diap**

GETTRUP E 1962 Thoracic proprioceptors in the flight system of locusts. *Nature 193: 498-499* **SO flight SR propr**

GETTRUP E 1963 Information carried by single impulses from a unicellular stretch receptor. *Acta physiol. Scand. 53, Suppl. 213: 46* **SO flight SR propr**

GETTRUP E 1963 Phasic stimulation of a thoracic stretch receptor in locusts. *J. exp. Biol. 40: 323-334* **SO flight SR propor**

GETTRUP E 1965 Sensory mechanisms in locomotion: the campaniform sensilla of the insect wing and their function during flight. *Cold Spring Harbor Symp. Quant. Biol. 30: 615-621* **SO mech flight**

GETTRUP E 1966 Sensory regulation of wing twisting in locusts. *J. exp. Biol. 44: 1-17* **SO mech flight**

GETTRUP E 1970 Activity in the locust nerve cord in response to wing-nerve stimulation. *J. exp. Biol. 52: 667-674* **SO flight**

GETTRUP E WILSON DM 1964 The lift-control reaction of flying locusts. *J. exp. Biol. 41: 183-190* **SO flight aero**

GEWECKE M 1970 Antennae: another wind-sensitive receptor in locusts. *Nature 225: 1263-1264* **flight windh SO mech ant**

GEWECKE M 1972 Antennen und Stirn-Scheitelhaare von *Locusta migratoria* L. als Luftströmungs-Sinnesorgan bei der Flugsteuerung. *J. comp. Physiol. 80: 57-94* **flight windh SO mech ant**

GEWECKE M 1972 Die Regelung der Fluggeschwindigkeit bei Heuschrecken und ihre Bedeutung für die Wanderflüge. *Verh. dt. zool. Ges. 1972: 247-250* **windh SO mech ant flight**

GEWECKE M 1972 Bewegungsmechanismus und Gelenkrezeptoren der Antennen von *Locusta migratoria* (L.) (Insecta, Orthoptera). *Z. Morphol. Tiere 71: 128-149* **flight windh SO mech ant anat**

GEWECKE M 1974 The antennae of insects as air-current sense organs and their relationship to the control of flight. *In: Barton Browne L (ed.), Experimental analysis of insect behaviour. Berlin, Heidelberg, New York: Springer Verlag. 100-113* **flight windh SO mech ant**

GEWECKE M 1975 The influence of the air-current organs on the flight behaviour of *Locusta migratoria. J. comp. Physiol. A 103: 79-95* **SO flight windh mech ant**

GEWECKE M 1977 Control of flight in relation to the air in *Locusta migratoria* (Insecta, Orthoptera). *J. Physiol. (Paris) 73: 581-592* **flight windh SO mech ant**

GEWECKE M 1979 Central projection of antennal afferents for the flight motor in *Locusta migratoria* (Orthoptera: Acrididae). *Entomologia generalis 5: 317-320* **SO flight windh mech ant**

GEWECKE M 1983 Comparative investigations of locust flight in the field and in the laboratory. *In: Nachtigall W (ed.), BIONA-report 2 "Insekten Flug"; Akad. Wiss. Mainz. Stuttgart & New York: Gustav Fischer Verlag. 11-20* **flight behav**

GEWECKE M HEINZEL H-G 1980 Aerodynamic and mechanical properties of the antennae as air-current sense organs in *Locusta migratoria.* I. Static characteristics. *J. comp. Physiol. A 139: 357-366* **SO flight ant**

GEWECKE M KIRSCHFELD K FEILER R 1989 Identification of optic lobe neurons of locusts by video films. *Biol. Cybern. 63: 411-420* **vis opt IN tech anat**

GEWECKE M KUTSCH W 1979 Development of flight behaviour in maturing adults of *Locusta migratoria*: I. Flight performance and wing-stroke parameters. *J. Insect Physiol. 25: 249-253* **flight aero devel rhythm**

GEWECKE M LEMMNITZ G 1989 Analysis of visual interneurones in locusts using video film and computer program for stimulation. *In: Erber J Menzel Pflüger H-J Todt D (eds.), Neural mechanisms of behavior: Proc. 2nd Internat. Congr. Neuroethology. Stuttgart & New York: Thieme Verlag p. 178* **vis IN brain tech**

GEWECKE M PHILIPPEN J 1978 Control of the horizontal flight-course by air-current sense organs in *Locusta migratoria. Physiol. Entomol. 3: 43-52* **SO flight windh mech ant**

GEWECKE M SAAGER F 1985 Die Heuschreckenantenne als Beschleunigungs-Sinnesorgan. *Verh. dt. zool. Ges. 1985: 240* **SO flight ant**

GIFFORD AN NICHOLSON RA PITMAN RM 1991 The dopamine and 5-hydroxytryptamine content of locust and cockroach salivary neurons. *J. exp. Biol. 161: 405-414* **saliv TS 5HT dopa dict MN**

GILES D USHERWOOD PNR 1985 The effects of putative aminoacid neurotransmitters on somata isolated from neurons of the locust central nervous system. *Comp. Biochem. Physiol. C 80: 231-236* **TS glu cellCult**

GILES DP JOY RT USHERWOOD PNR 1977 Growth of isolated locust neurones in culture. *J. Physiol. (Lond.) 276: 74P* **cellCult tech**

GILES DP USHERWOOD PNR 1985 Locust nymphal neurones in culture: a new technique for studying the physiology and pharmacology of insect central neurones. *Comp. Biochem. Physiol. C 80: 53-59* **cellCult tech**

GILLETT SD 1983 Primer pheromone and polymorphism in the desert locust. *Anim. Behav. 31: 221-230* **pher phase ant gut rev**

GIRARDIE A 1962 Rôles de la pars intercerebralis chez *Locusta migratoria. C. R. Acad. Sci. Paris Ser. D 254: 2269-2671* **brain NSC**

GIRARDIE A 1963 Functions of the locust pars intercerebralis. *Proc. nat. Acad. Sci. USA 4: 117-118* **brain NSC**

GIRARDIE A 1964 Fonction gonadotrope de la pars intercerebralis chez le mâle de *Locusta migratoria* (Orthoptère). *C. R. Acad. Sci. Paris Ser. D 258: 2910-2911* **brain NSC genit**

GIRARDIE A 1964 Pars intercerebralis and development in locust. *J. Insect Physiol. 10: 599-609* **brain NSC devel**

GIRARDIE A 1965 Réduction de l'intermue chez *Locusta migratoria* par implantation de pars intercerebralis. *C. R. Acad. Sci. Paris Ser. D 261: 4247-4249* **moult NSC brain**

GIRARDIE A 1965 Contribution à la étude de contrôle de l'activité des corpora allata par la pars intercerebralis chez *Locusta migratoria* (L.). *C. R. Acad. Sci. Paris Ser. D 261: 4876-4878* **pep retroc CAll NSC**

GIRARDIE A 1966 Contrôle de l'activité génitale chez *Locusta migratoria.* Mise en évidence d'un facteur gonadotrope et d'un facteur allatotrope dans la pars intercerebralis. *Bull. Soc. Zool. Fr. 91: 423-439* **brain NSC genit CAll retroc**

GIRARDIE A 1967 Contrôle neuro-hormonal de la métamorphose et de la pigmentation chez *Locusta migratoria cinerascens. Bull. Biol. Fr. Belg. 101: 79-114* **brain hormX col CAll retroc**

GIRARDIE A 1970 Neurosécrétions cérébrales chez les acridiens. *Bull. Soc. Zool. Fr. 95: 783-802* **NSC brain**

GIRARDIE A 1970 Mise en évidence dans le protocérébron de *Locusta migratoria* et de *Schistocerca gregaria* de nouvelles cellules neurosécrétrices contrôlant le métabolisme hydrique. *C. R. Acad. Sci. Paris Ser. D 271: 504-507* **brain ADH anat NSC**

GIRARDIE A 1971 Hormones et mécanismes endocrines contrôlant l'activité génitale de *Locusta migratoria. Archs. Zool. exp. gén. 112: 635-648* **retroc CC genit**

GIRARDIE A 1974 Recherches sur le rôle physiologique des cellules neurosécrétrices latérales du protocérébron de *Locusta migratoria migratorioides. Zool. Jb. (Physiol.) 78: 310-326* **pep retroc CAll NSC**

GIRARDIE A 1980 Apercu des corrélations neuro-humorales des insectes acridiens (Orthoptères). *Bull. Soc. Sc. Bret. 51: 29-51* **pep**

GIRARDIE A 1980 Contrôle neurosécrétoire du développement post-embryonnaire des insectes. *Bull. Soc. Zool. Fr. 105: 33-48* **pep devel**

GIRARDIE A 1983 Neurosecretion and reproduction in insects. *In: Downer RGH Laufer H (eds.): Invertebrate Endocrinology, Vol. 1: Endocrinology of insects. New York: Alan R. Liss Inc. 305-317* **pep behav**

GIRARDIE A 1990 Régulation endocrinienne du développement, de la reproduction et du polymorphisme phasaire des criquets grégarisables. *In: Aupelf-Uref J (ed.): La lutte anti-acridienne. Paris: John Libbey Eurotext. 119-127* **rev devel phase hormX**

GIRARDIE A BENTZ F 1969 Action de la pars intercerebralis et des corpora allata sur les protéines ovariennes de *Locusta migratoria migratorioides* (Orthoptère) au cours de la vitellogenèse. *C. R. Acad. Sci. Paris Ser. D 269: 2014-2017* **brain retroc CAll NSC**

GIRARDIE A CAZAL M 1965 Rôle de la pars intercerebralis et des corpora cardiaca sur la mélanisation chez *Locusta migratoria* (L.). *C. R. Acad. Sci. Paris Ser. D 261: 4525-4527* **NSC retroc col CC**

GIRARDIE A CAZAL M 1972 Contrôle endocrine des contractions de l'oviducte isolé de *Locusta migratoria. C. R. Acad. Sci. Paris Ser. D 274: 2208-2210* **hormX genit**

GIRARDIE A DE REGGI M 1978 Moulting and ecdysone release in response to electrical stimulation of protocerebral neurosecretory cells in *Locusta migratoria. J. Insect Physiol. 24: 797-802* **NSC moult brainStim**

GIRARDIE A GIRARDIE J 1966 Mise en évidence d'une activité neurosécrétrice des cellules C de la pars intercerebralis de *Locusta migratoria* L. par étude comparative histologique et ultrastructurale. *C. R. Acad. Sci. Paris Ser. D 263: 1119-1122* **NSC brain fineStruc anat**

GIRARDIE A GIRARDIE J 1967 Etude histochimique, histologique et ultrastructurale de la pars intercerebralis chez *Locusta migratoria* (Orthoptère). *Z. Zellforsch. mikr. Anat. 78: 54-75* **NSC brain fineStruc anat**

GIRARDIE A GIRARDIE J 1970 Mise en évidence par électrophorèse, de trois fractions protéiniques fuchsinophiles dans la pars intercerebralis de *Locusta migratoria. J. Insect Physiol. 16: 1745-1756* **NSC brain hormX**

GIRARDIE A GIRARDIE J 1972 Les cellules neurosécrétrices de la zone sous-ocellaire de *Locusta migratoria migratorioides*, R. et F.: étude histologique, cytologique, autoradiographique et expérimentale. *Acrida 1: 205-222* **oc NSC brain DH**

GIRARDIE A GIRARDIE J LAVENSEAU L PROUX J REMY C 1985 Insect neurosecretion: similarities and differences with vertebrates neurosecretion. *In: Kobayashi H Bern HA Urano A (eds.): Neurosecretion and the biology of neuropeptides. Tokyo: Japan Sci. Soc. Press/ Berlin, Heidelberg, New York: Springer Verlag. 392-400* **NSC**

GIRARDIE A GRANIER S 1973 Système endocrine et physiologie de la diapause imaginale chez le criquet égyptien, *Anacridium aegyptium*. *J. Insect Physiol. 19: 2341-2358* **pep retroc NSC diap**

GIRARDIE A JOLY P 1967 Mécanisme physiologique del'effet de groupe chez les acridiens. *Colloq. int. CNRS 173: 127-145* **hormX phase**

GIRARDIE A VOGEL A 1966 Etude de contrôle neurohormonal de l'activité sexuelle mâle de *Locusta migratoria* (L.). *C. R. Acad. Sci. Paris Ser. D 263: 543-546* **pep retroc NSC genit behav**

GIRARDIE J 1973 Aspects histologique, histochimique et ultrastructural de péricaryons neurosécréteurs latéreaux de protocérébron de *Locusta migratoria migratorioides* (Insecte: Orthoptère). *Z. Zellforsch. mikr. Anat. 141: 75-91* **NSC fineStruc brain**

GIRARDIE J 1975 Recherche en microscopie photonique et électronique des éléments neurosécréteurs tritocérébraux de *Locusta migratoria* (Insecte: Orthoptère). *Archs. Anat. micr. Morphol. exp. 64: 223-246* **NSC brain fineStruc**

GIRARDIE J 1977 Dynamique, après axotomie, des cellules neurosécrétrices protocérébrales médianes A1 & A2 du criquet migrateur. II. Etude ultrastructurale. *Archs. Anat. micr. Morphol. exp. 66: 305-322* **NSC fineStruc brain**

GIRARDIE J 1988 Isolement des neurohormones d'insectes: techniques, résultats et perspectives. *Ann. Soc. entomol. Fr. N.S. 24: 383-389* **pep tech**

GIRARDIE J 1990 Structure des glandes endocrines et chimie des hormones des criquets grégarisables. *In: Aupelf-Uref J (ed.): La lutte anti-acridienne. Paris: John Libbey Eurotext. 101-117* **fineStruc retroc pep**

GIRARDIE J ALLARD M FADDOUL A NEUZIL E 1983 Separation of the proteins of locust neurosecretory endings with HPLC. *J. Liqu. Chromatogr. 6: 2501-2512* **pep tech NSC**

GIRARDIE J BOUREME D COUILLARD F TAMARELLE M GIRARDIE A 1987 Anti-juvenile effect of neuroparsin A, a neuroprotein isolated from locust corpora cardiaca. *Insect Biochem. 17: 977-983* **pep neuropars retroc CC**

GIRARDIE J BOUREME D GIRARDIE A 1987 Productions sites of the three neurosecretory proteins characterized in the corpora cardiaca of the migratory locust. *Insect Biochem. 17: 29-36* **pep retroc neuropars brain NSC CC**

GIRARDIE J BOUREME D TAMARELLE M GIRARDIE A 1986 Three neurosecretory proteins in the nervous corpora cardiaca of the migratory locust. *Bull. Soc. zool. Fr. 111: 38-39* **NSC retroc CC hormX**

GIRARDIE J FADDOUL A GIRARDIE A 1985 Characterization of three neurosecretory proteins from A median neurosecretory cells of *Locusta migratoria* by coupled chromatographic, electrophoretic and isoelectrofocussing methods. *Insect Biochem. 15: 85-92* **pep NSC brain**

GIRARDIE J GIRARDIE A 1972 Evolution de la radioactivité des cellules neurosécrétrices de la pars intercerebralis chez *Locusta migratoria migratorioides* (Insecte Orthoptère) après injection de cystéine ^{35}S. Etude radiographique aux microscopes photoniques et électroniques. *Z. Zellforsch. mikr. Anat. 128: 212-226* **retroc NSC fineStruc brain**

GIRARDIE J GIRARDIE A 1977 Dynamique, après axotomie, des cellules neurosécrétrices protocérébrales médianes A1 & A2 du criquet migrateur. I. Etude histophysiologique. *Archs. Anat. micr. Morphol. exp. 66: 291-304* **retroc NSC anat brain**

GIRARDIE J GIRARDIE A 1977 Neurosécrétion médiane après section des nerfs hypocérébrocardiaques chez le criquet migrateur. *C. R. Acad. Sci. Paris Ser. D 284: 1433-1436* **NSC SGG brain**

GIRARDIE J GIRARDIE A 1977 Section des nerfs paracardiaques internes et libération provoquée du facteur gonadotrope médian chez le criquet migrateur. *Experientia 33: 1250-1251* **NSC brain genit**

GIRARDIE J GIRARDIE A 1977 "Last in, first out" in locust neurosecretory lobes of the corpus cardiaca. *Gen. comp. Endocrinol. 34: No. 1 (Abstracts): 112* **NSC CC pep retroc**

GIRARDIE J GIRARDIE A 1977 Libération provoquée in vitro du produit de neurosécrétion des cellules protocérébrales médianes chez le criquet migrateur. *J. Physiol. (Paris) 73: 707-721* **NSC pep brain**

GIRARDIE J GIRARDIE A HUET JC PERNOLLET JC 1988 Aminoacid sequence of locust neuroparsins. *FEBS Lett. 245: 4-8* **pep neuropars**

GIRARDIE J GIRARDIE A MOULINS M 1975 Preuves radiochimiques et physiologiques de l'activation par électrostimulation des cellules neurosécrétrices de la pars intercerebralis chez *Locusta migratoria. Gen. comp. Endocrinol. 25: 416-424* **NSC brainStim**

GIRARDIE J GIRARDIE A MOULINS M 1976 Etude radiochimique après électrostimulation de la dynamique fonctionelle des cellules neurosécrétrices protocérébrales médianes de *Locusta migratoria. Gen. comp. Endocrinol. 30: 410-418* **NSC brainStim**

GIRARDIE J HUET JC PERNOLLET JC 1990 The locust neuroparsins A: sequence and similarities with vertebrate and insect polypeptide hormones. *Insect Biochem. 20: 659-666* **pep neuropars**

GIRARDIE J HUET JC PERNOLLET JC GIRARDIE A 1990 Sequencing and biological activities of neuroparsins isolated from copora cardiaca of *Locusta migratoria. In: Borkovec AB Masler EP (eds.), Insect neurochemistry and neurophysiology. Clifton, N.J.: Humana Press Inc. 219-222* **pep neuropars**

GIRARDIE J MOULINS M GIRARDIE A 1974 Rupture de la diapause ovarienne d'*Anacridium aegyptium* par stimulation électrique des cellules neurosécrétrices protocérébrales médianes de la pars intercerebralis. *J. Insect Physiol. 20: 2261-2275* **NSC diap brainStim**

GIRARDIE J PERNOLLET JC HUET JC GIRARDIE A 1990 Isolation and structural characterisation of neuroparsins, polytropic neurohormones of the african locust. *In: McCaffery A Wilson I (eds.): Chromatography and isolation of insect hormones and pheromones. New York: Plenum Press. 223-230* **pep neuropars**

GIRARDIE J RÉMY C 1980 Particularités histo-cytologiques des prolongements distaux des deux cellules à "vasopressin-neurophysin-like" du criquet migrateur. *J. Physiol. (Paris) 76: 265-271* **anat suboes brain**

GIRARDIE J ROSSI C 1978 Preuves histologiques et ultrastructurales de deux catégories de cellules neurosécrétrices protocérébrales médianes de type A chez le criquet migrateur. *C. R. Acad. Sci. Paris Ser. D 286: 97-100* **NSC brain**

GIRARDIE J TOBE SS GIRARDIE A 1982 Electrostimulation of medial and lateral neurosecretory cells enhances juvenile hormone biosynthesis in locusts. *Gen. comp. Endocrinol. 46: 379-380* **JH brainStim NSC retroc CAll**

GISZTER SF 1985 Neural oscillators: pattern generators and pattern interpreters. *Soc. Neurosci. Abstr. 11: 477* **rhythm**

GIULIO L LUCARONI A 1967 "Test flash" interaction in the electroretinographic response of the compound eye. *Experientia 23: 524-543.* **ret**

GLINKA AV PSHENNIKOVA ES 1990 The role of juvenile hormone and a brain factor in the regulation of vitellogenin biosynthesis in the locust (*Locusta migratoria*). *Biokhimiya 55: 1445-1450. See also Biochemistry-English Transl. 55: 1077-1081* **hormX JH**

GODDEN DH 1970 The neural basis for locust jumping. *Am. Zool 9: 1139* **jump MN**

GODDEN DH 1975 The neural basis for locust jumping. *Comp. Biochem. Physiol. A 51: 351* **jump MN**

GOKULDAS M HUNT PA CANDY DJ 1988 The inhibition of lipid synthesis in vitro in the locust *Schistocerca gregaria* by factors from the corpora cardiaca. *Physiol. Entomol. 13: 43-48* **metab retroc CC lip inhib**

GOLDSWORTHY GJ 1969 Hyperglycaemic factors from the corpus cardiacum of *Locusta migratoria. J. Insect Physiol. 15: 2131-2140* **pep retroc AKH CC**

GOLDSWORTHY CJ 1983 The endocrine control of flight metabolism in locusts. *Adv. Insect Physiol. 17: 149-204* **AKH flight pep retroc rev**

GOLDSWORTHY GJ 1990 Hormonal control of flight metabolism in locusts. *In: Chapman RF Joern A (eds.), Biology of grasshoppers. New York: J. Wiley & Sons Inc. 205-226* **flight retroc CC NSC rev**

GOLDSWORTHY GJ JOHNSON RA MORDUE W 1972 In vivo studies on the release of hormones from the corpora cardiaca of locusts. *J. comp. Physiol. 79: 85-96* **pep retroc AKH CC**

GOLDSWORTHY GJ MALLISON K WHEELER CH 1986 The relative potencies of two known locust adipokinetic hormones. *J. Insect Physiol. 32: 145-157* **pep AKH**

GOLDSWORTHY GJ MALLISON K WHEELER CH GADE G 1986 Relative adipokinetic activities of members of the adipokinetic hormone/red pigment concentrating hormone family. *J. Insect Physiol. 32: 433-438* **pep AKH CPCH**

GOLDSWORTHY GJ WHEELER CH 1986 The endocrine control of flight metabolism in locusts. *In: Danthanarayana E (ed.), Insect flight: dispersal and migration. Berlin, Heidelberg, New York: Springer Verlag. 49-59* **flight TS oct pep**

GOLDSWORTHY GJ WHEELER CH 1989 Physiological and structural aspects of adipokinetic hormone function in locusts. *Pest. Sci. 25: 85-96* **pep retroc AKH rev**

GOLTZENÉ F PORTE A 1978 Endocrine control by neurosecretory cells of the pars intercerebralis and the corpora allata during the earlier phases of vitellogenesis in *Locusta migratoria migratorioides* R. & F. (Orthoptera). *Gen. comp. Endocrinol. 35: 35-45* **brain NSC retroc CAll**

GOODMAN CS 1974 Anatomy of locust ocellar interneurons: constancy and variability. *J. comp. Physiol. A 95: 185-201* **oc anat IN brain**

GOODMAN CS 1976 Anatomy of the ocellar interneurons of acridic grasshoppers. I. The large interneurons. *Cell Tiss. Res. 175: 166-183* **oc anat IN brain**

GOODMAN CS 1976 Constancy and uniqueness in a large population of small interneurons. *Science 193: 502-504* **oc anat IN brain DN**

GOODMAN CS 1977 Isogenic locusts and genetic variability in the effects of temperature on neuronal threshold. *J. comp. Physiol. A 117: 183-207* **MN thor var genet temp**

GOODMAN CS 1977 Neuron duplications and deletions in locust clones and clutches. *Science 197: 1384-1386* **oc anat IN brain**

GOODMAN CS 1978 Isogenic grasshoppers: genetic variability in the morphology of identified neurons. *J. comp. Neurol. 182: 681-706* **oc anat IN brain var genet**

GOODMAN CS 1979 Isogenic grasshoppers: genetic variability and development of identified neurons. *In: Breakefield E (ed.) Neurogenetics: genetic approaches to the nervous system. Amsterdam: Elsevier/ North-Holland Biomedical Press. 101-151* **oc IN devel anat genet**

GOODMAN CS 1982 Embryonic development of identified neurons in the grasshopper. *In: Spitzer NC (ed.), Neuronal development. New York: Plenum Press. 171-212* **devel IN thor**

GOODMAN CS 1984 Landmarks and labels that help developing neurons find their way. *Bioscience 34: 300-307* **devel rev**

GOODMAN CS BASTIANI MJ 1984 How embryonic nerve cells recognize one another. *Sci. Am. 251: 58-66* **devel rev**

GOODMAN CS BASTIANI MJ DOE CQ DU LAC S HELFLAND SL KUWADA JY THOMAS JB 1984 Cell recognition during neuronal development. *Science 225: 1271-1279* **devel**

GOODMAN CS BASTIANI MJ DOE CQ DULAC S 1986 Growth cone guidance and cell recognition in insect embryos. *In: Steinberg M (ed.): Developmental biology, Vol. 3: The cell surface in development and cancer. New York: Plenum Press. 283-300* **devel rev**

GOODMAN CS BASTIANI MJ RAPER JA THOMAS JB 1985 Cell recognition during neuronal development in grasshopper and *Drosophila* . *In: Edelman GM Gall WE Cowan WM (eds.), Molecular bases of neuronal development. Washington: Neuroscience Research Foundation. 295-316* **devel rev**

GOODMAN CS BATE CM 1981 Neuronal development in the grasshopper. *Trends in Neurosci. 4: 163-169* **devel anat devel**

GOODMAN CS BATE M SPITZER NC 1981 Embryonic development of identified neurons: origin and transformation of the H cell. *J. Neurosci. 1: 94-102* **devel DUM thor**

GOODMAN CS HEITLER WJ 1979 Electrical properties of spiking and non-spiking insect somata: normal, axotomized, and colchicine-treated neurons. *J. exp. Biol. 83: 95-121* **MN DUM thor**

GOODMAN CS HO RK BALL EE 1982 Pathfinding by growth cones in grasshopper embryos: guidance cues in the ectoderm and mesoderm. *In: Goodman CS Pearson KG (eds.), Neuronal development: cellular approaches in invertebrates. Neurosci. Res. Program Bull. 20: 847-859. Cambridge, Mass: MIT Press.* **devel**

GOODMAN CS O'SHEA M CAMAN R SPITZER NC 1979 Embryonic development of identified neurons: temporal pattern of morphological and biochemical differentiation. *Science 204: 1219-1222* **devel anat DUM IN thor**

GOODMAN CS PEARSON KG HEITLER WJ 1979 Variability of identified neurons in grasshoppers. *Comp. Biochem. Physiol. A 64: 455-462* **anat IN DUM var**

GOODMAN CS PEARSON KG SPITZER NC 1980 Electrical excitability: a spectrum of properties in the progeny of a single embryonic neuroblast. *Proc. nat. Acad. Sci. USA 77: 1676-1680* **DUM devel TS ACh nonspik local IN thor**

GOODMAN CS RAPER JA CHANG S HO R 1982 Grasshopper growth cones: divergent choices and labeled pathways. *Prog. Brain Res. 58: 283-304* **devel rev**

GOODMAN CS RAPER JA HO R CHANG S 1982 Pathfinding by neuronal growth cones in grasshopper embryos. *In: Developmental order: its origins and regulation. Symp. Soc. Dev. Biol. 40. New York: Alan Liss. 275-316* **devel**

GOODMAN CS RIDGE KA 1980 Development of an identified neuron after removal of its peripheral targets in grasshopper embryos cultured in vitro. *Soc. Neurosci. Abstr. 6: 496* **devel**

GOODMAN CS SHANKLAND M BENTLEY D 1983 Manipulation of the development of identified neurons in grasshopper embryos raised in culture. *In: Pfeiffer S (ed.): Neuroscience approached through cell culture, Vol. 2. Cleveland, Ohio: CRC Press. 79-88* **devel tech cellcult**

GOODMAN CS SPITZER NC 1979 Embryonic development of identified neurones: differentiation from neuroblast to neurone. *Nature 280: 208-214* **devel DUM IN thor**

GOODMAN CS SPITZER NC 1980 Embryonic development of neurotransmitter receptors in grasshoppers. *In: Sattelle DB Hall LM Hildebrand JG (eds.), Receptors for neurotransmitters, hormones and pheromones in insects. Amsterdam: Elsevier/North-Holland Biomedical Press. 195-207* **devel TS recept**

GOODMAN CS SPITZER NC 1981 The mature electrical properties of identified neurones in grasshopper embryos. *J. Physiol. (Lond.) 313: 369-384* **devel DUM thor**

GOODMAN CS SPITZER NC 1981 The development of electrical properties of identified neurons in grasshopper embryos. *J. Physiol. (Lond.) 313: 385-413* **devel DUM thor**

GOODMAN CS WILLIAMS JLD 1976 Anatomy of locust ocellar interneurons. II. The small interneurons. *Cell Tiss. Res. 175: 203-225* **oc anat IN brain**

GOODMAN LJ 1959 Hair receptors in locusts. Hair plates on the first cervical sclerites of the Orthoptera. *Nature 183: 1106-1107* **SO mech neck propr**

GOODMAN LJ 1965 The role of certain optomotor reactions in regulating stability in the rolling plane during flight in the desert locust, *Schistocerca gregaria. J. exp. Biol. 42: 385-408* **flight OM behav**

GOODMAN LJ 1975 The neural organization and physiology of the insect dorsal ocellus. *In: Horridge GA (ed.), The compound eye and vision in insects. Oxford: Clarendon Press. 515-548* **oc anat rev**

GOODMAN LJ 1981 Organization and physiology of the insect dorsal ocellar system. *In: Autrum H (ed.), Handbook of sensory physiology, Vol. VII/6C: Invertebrate visual centres and behaviour. Berlin, Heidelberg, New York: Springer Verlag. 201-286* **oc IN brain anat rev**

GOODMAN LJ MOBBS PG GUY RG 1977 Information processing along the course of a visual interneuron. *Experientia 33: 748-750* **oc IN brain anat**

GOODMAN LJ MOBBS PG KIRKHAM JB 1979 The fine structure of the ocelli of *Schistocerca gregaria*: the neural organization of the synaptic plexus. *Cell Tiss. Res. 196: 487-510* **oc brain fineStruc**

GOODMAN LJ PATTERSON JA MOBBS PG 1975 The projection of ocellar neurons within the brain of the locust *Schistocerca gregaria*. *Cell Tiss. Res. 157: 467-492* **oc IN brain anat**

GOOSEY MW CANDY DJ 1980 Effects of D- and L-octopamine and of pharmacological agents on the metabolism of locust flight muscle. *Biochem. Soc. Trans. 8: 532-533* **TS oct metab musc flight**

GOOSEY MW CANDY DJ 1980 The D-octopamine content of the haemolymph of the locust *Schistocerca americana gregaria* and its elevation during flight. *Insect Biochem. 10: 393-397* **flight oct blood**

GOOSEY MW CANDY DJ 1982 The regulation of substrate oxidation in locust flight muscle by a factor from the corpus cardiacum. *Biochem. Soc. Trans. 10: 276* **metab musc flight retroc CC hormX**

GOOSEY MW CANDY DJ 1982 The release and removal of octopamine by tissues of the locust *Schistocerca americana gregaria*. *Insect Biochem. 12: 681-685* **oct blood**

GORELKIN VS KARELIN YA SVIDERSKY VL 1990 Proprioceptive control of flight in insects. *In: Gribakin FG Wiese K Popov AV (eds.), Sensory systems and communication in arthropods. Basel, Boston, Berlin: Birkhäuser Verlag. 115-120* **windh SO mech anat suboes thor CO wing**

GORELKIN VS SVIDERSKY VL DAVID OF 1981 [Functional characteristics of receptors of chordotonal organ in the fore wings of the locust, *Locusta migratoria]*. (in Russian.). *Zh. evol. Biochim. Fiziol. 17: 259-263* **CO flight SO propr**

GÖRING C 1990 Physiology and morphology of flight interneurones in the suboesophageal ganglion of *Locusta migratoria*. *In: Elsner N Roth G (eds.) Brain - Perception - Cognition: Proc. 18th Göttingen Neurobiol. Conf. Stuttgart, New York: Thieme Verlag. Abstract 71.* **flight IN anat suboes**

GOURANTON J 1964 Contribution à l'étude de la structure des ganglions cérébrioides de *Locusta migratoria migratorioides*. *Bull. Soc. Zool. Fr. 89: 785-797* **anat brain**

GOURDOUX L MOREAU R BEN KHAY A DUTRIEU J 1985 In vivo glucose catabolism in male adult *Locusta migratoria*: effects of corpora cardiaca. *Comp. Biochem. Physiol. B 81: 485-489* **retroc CC metab**

GRABER V 1871 Die tympanalen Sinnesapparate der Orthopteren. *Denkschr. Akad. Wiss. Wien 36: 1-140* **SO tymp anat aud**

GRAMOLL S 1988 Activity of metathoracic interneurons during stridulation in the acridid grasshopper *Omocestus viridulus* L. *J. comp. Physiol. A 163: 813-826* **strid IN thor**

GRAMOLL S ELSNER N 1987 Morphology of local "stridulation" interneurons in the metathoracic ganglion of the acridid grasshopper *Omocestus viridulus* L. *J. comp. Neurol. 263: 593-606* **strid thor IN anat local**

GRAMOLL S ELSNER N 1987 "Stridulatory" interneurons in the metathoracic ganglion of the grasshopper *Omocestus viridulus* L. *Naturwiss. 74: 45* **strid thor IN**

GRANT GRM 1949 The sensory pits of insects considered as dielectric wave guides and resonators to infra-red rays. *Proc. R. Soc. Queensld. 60: 89-98* **SO therm**

GRATION KAF 1982 Activation of channels in locust muscle by amino acids. *In: Neuropharmacology of insects. London: Pitman (Ciba Foundation Symp. 88). 240-252* **rev musc TS glu NM channel**

GRATION KAF CLARK RB USHERWOOD PNR 1979 Three types of L-glutamate receptor on junctional membrane of locust muscle fbres. *Brain Res. 171: 360-364* **TS glu NM recept**

GRATION KAF LAMBERT JJ RAMSEY R USHERWOOD PNR 1981 Potency of glutamate and agonists determined by patch clamp analysis of individual receptors on locust muscle. *J. Physiol. (Lond.) 317: 34-35* **TS glu NM recept channel**

GRATION KAF LAMBERT JJ RAMSEY RL RAND PR USHERWOOD PNR 1981 Agonist potency determination by patch clamp analysis of single glutamate receptors. *Brain Res. 230: 400-405* **musc TS glu NM channel recept**

GRATION KAF LAMBERT JJ RAMSEY RL RAND PR USHERWOOD PNR 1982 Closure of membrane channels gated by glutamate receptors may be a two-step process. *Nature 295: 599-601* **musc TS glu NM channel recept**

GRATION KAF LAMBERT JJ RAMSEY RL USHERWOOD PNR 1980 Non-random openings and concentration-dependent life-times of glutamate-gated channels in muscle membranes. *Nature 291: 423-425* **musc TS glu NM channel**

GRATION KAF LAMBERT JJ USHERWOOD PNR 1981 Glutamate-activated channels in locust muscle. *In: Pethes G Frenyo VL (eds.): Advances in animal and comparative physiology. Oxford & New York: Pergamon Press. 377-383* **TS glu channel musc**

GRATION KAF USHERWOOD PNR 1980 Interactions of glutamate with amino acid receptors on locust muscle. *Verh. dt. zool. Ges. 1980: 122-132* **musc TS glu NM recept**

GRAY EG 1960 The fine structure of the insect ear. *Phil. Trans. R. Soc. Lond. (Biol.) 243: 75-94* **aud SO tymp anat fineStruc**

GRAY EG PUMPHREY RJ 1958 Ultrastructure of the insect ear. *Nature 181: 618.* **aud SO tymp fineStruc**

GREENWOOD M CHAPMAN RF 1984 Differences in numbers of sensilla on the antennae of solitarious and gregarious *Locusta migratoria* L. (Orthoptera: Acrididae). *Int. J. Insect Morphol. Embryol. 13: 295-301* **SO ant anat**

GRISS C ROWELL CHF 1986 Three descending interneurons reporting deviation from course in the locust. I. Anatomy. *J. comp. Physiol. A 158: 765-774* **DN flight anat OM**

GROSCH A CALLENDER F PETERSEN M COKL C KALMRING K 1985 Vibration receptors of larvae and of imagines in locusts: location on the legs, central projections and physiology. *In: Kalmring K Elsner N (eds.) Acoustic and vibrational communication in insects. Hamburg & New York: Paul Parey Verlag. 151-161* **SO leg vibr anat subg**

GROSSMAN HJ NORTHROP RB 1976 Tests of the limiting visual resolution of the DCMD system of the lubber grasshopper to rectangular gratings. *J. Insect Physiol. 22: 799-808* **DN vis rom**

GUTHRIE DM 1962 Control of the ventral diaphragm in an insect. *Nature 196:1010-1012* **VD**

GUTHRIE DM 1964 Observations on the flight apparatus in the locust *Schistocerca gregaria. Quart. J. micr. Sci. 104: 183-201* **flight windh mech anat SO**

GUTHRIE DM 1966 The function and fine structure of the cephalic airflow receptors in *Schistocerca gregaria. J. Cell Sci. 1: 463-470* **flight windh mech SO fineStruc**

GUY R GOODMAN LJ MOBBS P 1977 Ocellar connections with the ventral nerve cord in the locust, *Schistocerca gregaria*: electrical and anatomical characteristics. *J. comp. Physiol. A 115: 337-350* **oc DN anat**

GWILLIAM GF BURROWS M 1980 Electrical characteristics of the membrane of an identified insect motor neurone. *J. exp. Biol. 86: 49-62* **MN leg**

GYNTHER IC PEARSON KG 1986 Intracellular recordings from interneurones and motoneurones during bilateral kicks in the locust: implications for mechanisms controlling the jump. *J. exp. Biol. 122: 323-343* **jump kick MN IN**

GYNTHER IC PEARSON KG 1987 Evaluation of a hypothesis for the neuronal control of jumping in the locust. *Soc. Neurosci. Abstr. 13: 1059* **IN jump**

GYNTHER IC PEARSON KG 1989 An evaluation of the role of identified interneurons in triggering kicks and jumps in the locust. *J. Neurophysiol. 61: 45-57* **IN anat thor jump**

HACKNEY CM ALTMAN JS 1982 Cobalt mapping of the nervous system: how to avoid artifacts. *J. Neurobiol. 13: 403-411* **tech**

HAFEZ M ABOUL-NASR AE SHANBAKI N 1970 Comparative studies on the nervous system of two grasshoppers in Egypt, *Poecilocerus bufonius* Klug. and *Tmethis pulchripennis* Serville (Orthoptera - Acrididae). *Bull. Fac. Sci. Cairo Univ. 43: 157-188* **anat abd genit pyrg pamph**

HAFEZ M IBRAHIM MI 1960 Anatomical studies on *Acrida pellucida. Bull. Soc. entomol. Egypte 44: 451-476* **anat**

HAHNEL C BRÄUNIG P 1989 Dorsal, unpaired median (DUM) neurones of the locust suboesophageal ganglion (SOG). *In: Elsner N Singer W (eds.), Dynamics and plasticity in neuronal systems: Proc. 17th. Göttingen Neurobiology Conference. Stuttgart, New York: Thieme Verlag. Abstract 53* **DUM brain suboes anat retroc SGG**

HALASY K BENEDECZKY I 1984 Comparative ultrastructural study on the neuromuscular junction in the muscles of the alimentary canal of *Locusta migratoria* and *Helix pomatia. Acta biol. Szeged 127-141* **NM gut fineStruc**

HALASY K BENEDECZKY I CSOKNYA M 1988 Ultrastructural comparison of enteric neuromuscular junctions in some invertebrates. *In: Salánki J S-Rózsa K (eds.), Neurobiology of invertebrates. (Symp. Biol. Hung. 36). Budapest. 199-211* **gut NM fineStruc**

HALASY K BENEDECZKY I FEKETE E TOTH L GABRIEL R 1988 Enteric neuromuscular junctions: comparison of ultrastructural features in different phylogenetic groups. *Neurosci. 25: 147-162* **NM gut fineStruc**

HALE JP BURROWS M 1985 Innervation patterns of inhibitory motor neurones in the thorax of the locust. *J. exp. Biol. 117: 401-413* **MN inhib leg anat**

HALEX H BECKMAN EA KAISER W KALMRING K 1988 Strukturbreitenanalyse an Nervenfasern mit einem optischen Messverfahren. *Leitz Mitt. Wiss. u. Tech. 9: 95-98* **aud SO tech**

HALEX H KAISER W KALMRING K 1988 Projection areas and branching patterns of the tympanal receptor cells in migratory locusts, *Locusta migratoria* and *Schistocerca gregaria. Cell Tiss. Res. 253: 517-528* **aud anat thor SO tymp**

HALEX M WAGNER HJ KALMRING K 1988 Synaptic contacts of tympanic receptor cells in the thoracic ventral nerve cord of locusts. *In: Elsner N Barth FG (eds.), Sense organs: interfaces between environment and behaviour: Proc.16th. Göttingen Neurobiol. Conf. Stuttgart, New York: Thieme Verlag. Abstract 149* **aud SO tymp**

HALFMANN K ELSNER N 1978 Larval stridulation in acridid grasshoppers. *Naturwiss. 65: 265* **strid devel behav**

HAMILTON AG 1955 Parthenogenesis in the desert locust *Schistocerca gregaria* and its possible effect on the maintenance of the species. *Proc. R. entomol. Soc. Lond. (A) 30: 104-114* **genet tech**

HAMILTON HL 1939 Action of acetylcholine, atropine and nicotine on the heart of the grasshopper (*Melanoplus differentialis*). *J. cell. comp. Physiol. 13: 91-103* **heart TS ACH tox**

HANKE W BREER H 1987 A neuronal acetylcholine receptor in planar lipid bilayers. *J. gen. Physiol. 90: 855-879* **recept TS ACh tech**

HANRAHAN J PHILLIPS JE SPRING JH PROUX J THOMSON B 1985 Cellular actions of a neuropeptide hormone on ion transport across an insect renal epithelium. *Regulatory Peptides 4: 131-135* **pep ADH**

HANSTRÖM B 1940 Inkretorische Organe, Sinnesorgane und Nervensystem des Kopfes einiger niederer Insektenordnungen. *K. svenska Vetensk. Akad. Handl. 18: 266 pp.* **brain retroc anat**

HARLOW PA 1958 The action of drugs on the nervous system of the locust, *Locusta migratoria. Ann. appl. Biol. 46: 55-73* **tox**

HARRELSON AL BASTIANI MJ SNOW PM GOODMAN CS 1987 From cell ablation experiments to surface glycoproteins: Selective fasciculation and the search for axonal recognition molecules. *In: Easter SS Barald KF Carlson BM (eds.): From message to mind: frontiers in developmental neurobiology. Sunderland, Mass.: Sinauer Associates. 96-109* **devel**

HARRELSON AL GOODMAN CS 1988 Growth cone guidance in insects: fasciclin II is a member of the immunoglobulin superfamily. *Science 242: 700-708* **devel**

HARRINGTON NG 1955 X-ray-induced "stickiness" of the chromosomes of the *Chortophaga* neuroblast in relation to dose and mitotic stage at treatment. *Radiation Res. 2: 84-90.* **devel tox**

HARRISON JF 1989 Ventilatory frequency and haemolymph acid-base status during short-term hypercapnia in the locust, *Schistocerca nitens. J. Insect Physiol. 35: 809-814* **resp blood**

HARRISON JF WONG CJH PHILLIPS JE 1990 Haemolymph buffering in the locust, *Schistocerca gregaria. J. exp. Biol. 154: 573-579* **blood pH**

HARRISON JM 1988 Temperature effects on haemolymph acid-base status in vivo and in vitro in the two-striped grasshopper *Melanoplus bivittatus. J. exp. Biol. 140: 421-435* **temp pH tech blood**

HARRISON JM 1989 Temperature effects on intra- and extracellular acid-base status in the American locust, *Schistocerca nitens. J. comp. Physiol. B 158: 763-770* **blood temp**

HARTLINE HK 1928 A quantitative and descriptive study of the electrical response to illumination of the arthropod eye. *Am. J. Physiol. 83: 466-483* **ret**

HASKELL PT 1955 Intensité sonore des stridulations de quelques Orthoptères britanniques. *In: Busnel RG (ed.), Colloque sur l'acoustique des Orthoptères. Ann. Epiphyt. (fasc. hors sér.), 154-168* **aud strid**

HASKELL PT 1956 Hearing in certain Orthoptera. I. Physiology of sound receptors. *J. exp. Biol. 33: 756-766* **strid aud SO mech cerc abd tymp**

HASKELL PT 1956 Hearing in certain Orthoptera. II. The nature of the response of certain receptors to natural and imitation stridulation. *J. exp. Biol. 33: 767-776* **strid aud tymp cerc SO mech**

HASKELL PT 1957 The influence of flight noise on behaviour in the desert locust *Schistocerca gregaria* (Forsk.). *J. Insect Physiol. 1: 52-75* **flight aud behav**

HASKELL PT 1959 Function of certain prothoracic hair receptors on the desert locust. *Nature 183: 1107* **SO thor flight mech propr head neck**

HASKELL PT 1960 The sensory equipment of the migratory locust. *In: Lowenstein OE (ed.), Sensory specialisation in response to environmental demands. Symp. zool. Soc. London 3: 1-23* **SO mech propr vis neck flight rev**

HASKELL PT 1962 Sensory factors influencing phase change in locusts. *Colloq. int. CNRS 114: 145-163* **SO phase**

HASKELL PT MOORHOUSE JE 1963 A blood-borne factor influencing the activity of the central nervous system of the desert locust. *Nature 197: 56-58* **hormX phase**

HASKELL PT MORDUE AJ (LUNTZ) 1969 The role of mouthpart receptors in the feeding behaviour of *Schistocerca gregaria. Entomol. exp. appl. 12: 591-610* **mpt SO feed chem**

HASKELL PT SCHOONHOVEN LM 1969 The function of certain mouthpart receptors in relation to feeding in *Schistocerca gregaria* and *Locusta migratoria migratorioides. Entomol. exp. appl. 12: 423-440* **mpt SO chem feed**

HEALD LJ 1968 Interaction between ocellar and compound eye visual units in *Schistocerca gregaria. Proc. 13th. Int. Congr. Entomol., Moscow, Vol. II: 19* **oc vis IN**

HEATHCOTE RD 1980 Physiological development of a monosynaptic connection involved in an adult insect behaviour. *J. comp. Neurol. 191: 155-166* **MN SO SR flight thor devel**

HEATHCOTE RD 1980 Embryonic morphogenesis of an identified sensory neuron. *Soc. Neurosci. Abstr. 6: 495* **SR devel propr**

HEATHCOTE RD 1981 Differentiation of an identified sensory neuron (SR) and associated structures (CTO) in grasshopper embryos. *J. comp. Neurol. 202: 1-18* **SR CO devel SO propr**

HECKMANN R KUTSCH W 1990 Common neural "Bauplan" in Tracheata? Innervation of the dorsal longitudinal muscles. *In: Elsner N Roth G (eds.) Brain - Perception - Cognition: Proc. 18th Göttingen Neurobiol. Conf. Stuttgart, New York: Thieme Verlag. Abstract 39* **anat MN thor musc innerv homol**

HEDWIG B 1984 Activity and deafness of auditory interneurons during stridulation in the grasshopper *Omocestus viridulus* (L.). *Naturwiss. 71: 380-381* **aud strid IN thor**

HEDWIG B 1986 On the role in stridulation of plurisegmental interneurons of the acridid grasshopper *Omocestus viridulus.* I. Anatomy and physiology of descending cephalothoracic interneurons. *J. comp. Physiol. A 158: 413-427* **strid IN jump resp**

HEDWIG B 1986 On the role in stridulation of plurisegmental interneurons of the acridid grasshopper *Omocestus viridulus.* II. Anatomy and physiology of ascending and T-shaped interneurons. *J. comp. Physiol. A 158: 429-444* **strid aud IN jump cerc abd**

HEDWIG B 1988 Peripheral versus central influences on auditory information processing in flying locusts. *In: Elsner N Barth FG (eds.), Sense organs: interfaces between environment and behaviour: Proc. 16th. Göttingen Neurobiol. Conf. Stuttgart, New York: Thieme Verlag. Abstract 26* **aud SO mech IN thor flight**

HEDWIG B 1988 Activation and modulation of auditory receptors in *Locusta migratoria* by respiratory movements. *J. comp. Physiol. A 162: 237-246* **SO aud resp tymp mod**

HEDWIG B 1989 Stridulation modulates auditory information processing in acridid grasshoppers. *In: Elsner N Singer W (eds.), Dynamics and plasticity in neuronal systems: Proc. 17th. Göttingen Neurobiology Conference. Stuttgart, New York: Thieme Verlag. Abstract 148* **strid aud IN thor mod**

HEDWIG B 1989 Stridulation interferes with auditory information processing in acridid grasshoppers. *In: Erber J Menzel Pflüger H-J Todt D (eds.), Neural mechanisms of behavior: Proc. 2nd Internat. Congr. Neuroethology. Stuttgart & New 6York: Thieme Verlag. Abstract 66* **aud SO tymp IN thor**

HEDWIG B 1989 Modulation of auditory information processing in tethered flying locusts. *J. comp. Physiol. A 164: 409-422* **aud tymp IN SO flight thor**

HEDWIG B 1990 The intersegmental network underlying stridulation in the grasshopper *Omocestus viridulus* L. *In: Gribakin FG Wiese K Popov AV (eds.), Sensory systems and communication in arthropods. Basel, Boston, Berlin: Birkhäuser Verlag. 189-192* **IN thor strid**

HEDWIG B 1990 Modulation of auditory responsiveness in stridulating grasshoppers. *J. comp. Physiol. A 167: 847-856* **IN thor strid aud SO tymp mod mod**

HEDWIG B ELSNER N 1985 Sound production and sound detection in a stridulating acridid grasshopper (*Omocestus viridulus*). *In: Kalmring K Elsner N (eds.) Acoustic and vibrational communication in insects. Hamburg & New York: Paul Parey Verlag. 61-72* **strid aud facnonresp**

HEDWIG B ELSNER N 1985 Gemeinsame Struktur- u. Funktionsanalyse von Nervenzellen. *Leitz Mitt. Wiss. u. Tech. 8: 221-227* **anat IN**

HEDWIG B GRAMOLL S ELSNER N 1990 Stridulation interneurons in the metathoracic ganglion of the acridid grasshopper *Omocestus viridulus. In: Elsner N Roth G (eds.) Brain - Perception - Cognition: Proc. 18th Göttingen Neurobiol. Conf. Stuttgart, New York: Thieme Verlag. Abstract 116* **IN thor strid**

HEDWIG B LANG F ELSNER N 1987 Activation and modulation of locust auditory receptors by respiration and leg movement. *In: Elsner N Creutzfeld O (eds.), New frontiers in brain research: Proc. 15th Göttingen Neurobiol. Conf. Stuttgart, New York: Thieme Verlag. Abstract 84* **strid tymp resp aud SO**

HEDWIG B LANG F ELSNER N 1988 The interference of sound and movement stimuli in tympanal receptors of *Locusta migratoria. J. comp. Physiol. A 163: 243-252* **aud strid SO tymp**

HEDWIG B PEARSON KG 1984 Patterns of synaptic input to identified flight motoneurons in the locust. *J. comp. Physiol. A 154: 745-760* **MN flight thor anat**

HEDWIG B GRAMOLL S ELSNER N 1990 Stridulatory interneurons in the metathoracic ganglion of the grasshopper *Omocestus viridulus* L. *In: Gribakin FG Wiese K Popov AV (eds.), Sensory systems and communication in arthropods. Basel, Boston, Berlin: Birkhäuser Verlag. 193-198* **IN thor strid**

HEDWIG B LANG F ELSNER N 1990 Modulation of auditory information processing by grasshoppers and locusts. *In: Gribakin FG Wiese K Popov AV (eds.), Sensory systems and communication in arthropods. Basel, Boston, Berlin: Birkhäuser Verlag. 193-197* **IN mod thor strid**

HEERMA W VERSLUIS C LANKHOF H OUDEJANS RCHM KOOIMAN FP BEENAKKERS AMT 1991 Structure determination of adipokinetic hormone using fast atom bombardment tandem mass spectrometry. An unknown adipokinetic hormone (AKH-III) from *Locusta migratoria. Anal. Chim. Acta 248: 553-561* **pep AKH**

HEINZEL H-G 1983 Rezeption von Luftströmungen und ihre Bedeutung für den Flug der Wanderheuschrecke. *In: Nachtigall W (ed.), BIONA-report 2 "Insekten Flug"; Akad. Wiss. Mainz. Stuttgart & New York: Gustav Fischer Verlag. 53-69* **flight SO windh mech ant**

HEINZEL H-G GEWECKE M 1979 Directional sensitivity of the antennal campaniform sensilla in locusts. *Naturwiss. 66: 212-213* **flight ant SO mech**

HEINZEL H-G GEWECKE M 1987 Aerodynamic and mechanical properties of the antennae as air-current sense organs in *Locusta migratoria.* II. Dynamic characteristics. *J. comp. Physiol. A 161: 671-680* **flight SO mech ant propr rhythm**

HEITLER WJ 1974 The locust jump. Specialisations of the metathoracic femoral-tibial joint. *J. comp. Physiol. 89: 93-104* **jump leg anat**

HEITLER WJ 1977 The locust jump. III. Structural specialisations of the metathoracic tibiae. *J. exp. Biol. 67: 29-36* **jump leg anat**

HEITLER WJ 1983 Suppression of a locust visual interneurone (DCMD) during defensive kicking. *J. exp. Biol. 104: 203-216* **DN jump kick inhib**

HEITLER WJ BRÄUNIG P 1988 The role of fast extensor motor activity in the locust kick reconsidered. *J. exp. Biol. 136: 289-310* **MN thor jump kick**

HEITLER WJ BURROWS M 1977 The locust jump. I. The motor programme. *J. exp. Biol. 66: 203-219* **jump MN thor**

HEITLER WJ BURROWS M 1977 The locust jump. II. Neural circuits of the motor programme. *J. exp. Biol. 66: 221-241* **jump MN IN thor anat SO subg brun**

HEITLER WJ GOODMAN CS 1978 Multiple sites of spike initiation in a bifurcating locust neurone. *J. exp. Biol. 76: 63-84* **DUM thor**

HEITLER WJ GOODMAN CS ROWELL CHF 1977 The effects of temperature on the threshold of identified neurons in the locust. *J. comp. Physiol. A 117: 163-182* **MN IN thor temp**

HEKIMI S BURKHART W MOYER M FOWLER E O'SHEA M 1989 Dimer structure of a neuropeptide precursor established: consequences for processing. *Neuron 2: 1363-1368* **pep AKH**

HEKIMI S O'SHEA M 1987 Identification and purification of two precursors of the insect neuropeptide adipokinetic hormone. *J. Neurosci. 7: 2773-2783* **pep AKH**

HEKIMI S O'SHEA M 1989 Antisera against AKHs and AKH precursors for experimental studies of an insect neurosecretory system. *Insect Biochem. 19: 79-83* **pep AKH**

HEKIMI S O'SHEA M 1989 Biosynthesis of adipokinetic hormones (AKHs): further characterization of precursors and identification of novel products of processing. *J. Neurosci. 9: 996-1003* **pep AKH**

HENRY LM 1968 Musculature of the cervical region in insects. *Microentomology 23: 95-105* **neck musc SO mech anat**

HENSLER K 1987 Flight steering in locusts: parallel encoding of deviations from straight flight and head movements in the same deviation detector neuron and its functional significance. *In: Elsner N Creutzfeld O (eds.), New frontiers in brain research: Proc. 15th Göttingen Neurobiol. Conf. Stuttgart, New York: Thieme Verlag. Abstract 51* **flight DN propr vis OM**

HENSLER K 1988 Convergent processing of deviations from course, head position, and flight activity in descending neurons of *Locusta migratoria. In: Elsner N Barth FG (eds.) Sense organs: interfaces between environment and behaviour: Proc.16th. Göttingen Neurobiol. Conf. Stuttgart, New York: Thieme Verlag. Abstract 21* **DN flight vis propr wind OM**

HENSLER K 1988 The pars intercerebralis neurone PI(2)5 of locusts: convergent processing of inputs reporting head movements and deviations from straight flight. *J. exp. Biol. 140: 511-533* **flight propr vis DN anat OM**

HENSLER K 1989 Neural processing and behavioral consequences of head movements in locust flight. *In: Erber J Menzel Pflüger H-J Todt D (eds.), Neural mechanisms of behavior: Proc. 2nd Int. Congr. Neuroethology. Stuttgart & New York: Thieme Verlag. Abstract 1* **IN thor vis propr flight DN OM**

HENSLER K 1989 Corrective flight steering in locusts: convergence of extero- and proprioceptive inputs in descending deviation detectors. *In: Naresh Singh R Strausfeld NJ (eds.) Neurobiology of sensory systems. New York: Plenum Press. 531-554* **flight vis DN neck head propr OM**

HENSLER K 1990 Neural control of optomotor head rolling in locusts. *Naturwiss. 77: 35-37* **flight MN IN neck thor suboes anat OM**

HENSLER K ROBERT D 1990 Compensatory head rolling during corrective flight steering in locusts. *J. comp. Physiol. A 166: 685-693* **flight propr vis behav OM**

HENSLER K ROWELL CHF 1990 Control of optomotor responses by descending deviation detectors in intact flying locusts. *J. exp. Biol. 149: 191-205* **flight vis DN neck head OM**

HERAULT JP GIRARDIE J PROUX J 1985 Separation and characteristics of antidiuretic factors from the corpora cardiaca of the migratory locust. *Int. J. Invert. Reprod. Devel. 8: 325-335* **ADH pep CC retroc**

HERAULT JP GIRARDIE J PROUX J 1986 Study of antidiuretic factors from the corpora cardiaca of the migratory locust. *Bull. Soc. Zool. Fr. 111: 39-40* **pep ADH retroc CC**

HERAULT JP GIRARDIE J PROUX J 1988 Further characterization of the antidiuretic factor from the glandular part of the corpora cardiaca of the migratory locust. *Int. J. Invert. Reprod. Devel. 13: 183-192* **retroc ADH pep CC**

HERAULT JP PROUX P 1987 Cyclic AMP: the second messenger of an antidiuretic hormone from the glandular lobes of the migratory locust corpora cardiaca. *J. Insect Physiol. 33: 487-491* **cAMP retroc pep ADH CC**

HERHAERT P DE LOOF A HUYBRECHTS R DELANG I THEUNIS W CLOTTENS F SCHOOFS L DWINNEN K VANDESAND F 1986 A new alternative for simultaneous screening of 96 hybridoma clones for tissue-specific antibody production selects a monoclonal antibody to insect corpus cardiacum. *J. Neurosci. Meth. 17: 261-268* **retroc CC**

HERRMANN C PFLÜGER H-J 1988 Zur Entwicklung und Innervierung der Genitalmuskulatur bei der Wanderheuschrecke, *Locusta migratoria. In: Elsner N Barth FG (eds.), Sense organs: interfaces between environment and behaviour: Proc. 16th. Göttingen Neurobiol. Conf. Stuttgart, New York: Thieme Verlag. Abstract 102* **MN abd genit anat devel innerv musc**

HETRU C LI KW BULET P LAGUEUX M HOFFMANN JA 1991 Isolation and structural characterization of an insulin-related molecule, a predominant neuropeptide from *Locusta migratoria. Eur. J. Biochem. 201: 495-499* **pep insul retroc CC**

HEUKAMP U 1983 Die Rolle von Mechanorezeptoren im Flugsystem der Wanderheuschrecke (*Locusta migratoria* L.). *Dissertation., Universität Köln, Germany.* **flight SO mech**

HEUKAMP U 1984 Sensory regulation of the pleuroalar muscles in the migratory locust. *Naturwiss. 71: 481-482.* **MN flight SO windh mech teg propr**

HEUKAMP U WENDLER G 1981 Untersuchung der Bewertung sensorischer Eingänge, die auf Flugmotorneurone von Wanderheuschrecken konvergieren: eine neue Methode. *Verh. dt. zool. Ges. 1980: 236* **flight SO tech**

HIERS J 1928 Organe tympanale du *Schistocerca gregaria* Forsk. *Proc. 5th. Int. Locust Conf., Brussels 1928: 434-445* **aud SO tymp anat**

HIETTER H LUU B GOLTZENE F ZACHARY D HOFFMANN JA VAN DOSSELAER A 1989 Isolation and structure of two novel 6kD dimeric peptides from the corpora cardiaca of the insect *Locusta migratoria*: molecular mass determination by mass spectrometry. *Eur. J. Biochem. 182: 77-84* **retroc CC pep**

HIETTER H VAN DORSSELAER A GREEN B DENOROY L HOFFMANN JA LUU B 1990 Isolation and structure elucidation of a novel 5-kDa peptide from neurohaemal lobes of the corpora cardiaca of *Locusta migratoria* (Insecta, Orthoptera). *Eur. J. Biochem. 187: 241-247* **retroc pep CC**

HIETTER H VAN DOSSELAER A LUU B 1991 Characterization of three structurally-related 8 to 9 kD monomeric peptides present in the corpora cardiaca of *Locusta*: a revised structure for the neuroparsins. *Insect Biochem. 21: 259-264* **CC retroc neuropars pep**

HIGHNAM KC 1961 Induced changes in the amount of material in the neurosecretory system of the desert locust. *Nature 191: 199-200* **anat brain NSC retroc**

HIGHNAM KC 1961 The histology of the neurosecretory system of the adult female desert locust, *Schistocerca gregaria. Quart. J. micr. Sci. 102: 27-38* **anat brain NSC retroc**

HIGHNAM KC 1962 Neurosecretory control of ovarian development in *Schistocerca gregaria*, and its relation to phase differences. *In: Physiologie, comportement et écologie des acridiens en rapport avec la phase. Coll. Internat. CNRS 114: 107-121* **brain NSC phase**

HIGHNAM KC 1962 Variation in neurosecretory activity during oocyte development in *Schistocerca gregaria. J. Endocrinol. 24: 4-5* **brain NSC**

HIGHNAM KC 1962 Neurosecretory control of ovarian development in *Schistocerca gregaria. Quart. J. micr. Sci. 103: 57-72* **brain NSC retroc**

HIGHNAM KC HASKELL PT 1964 The endocrine systems of isolated and crowded *Locusta* and *Schistocerca* in relation to oocyte growth, and the effects of flying upon maturation. *J. Insect Physiol. 10: 849-864* **NSC brain flight**

HIGHNAM KC HILL L GINGELL DJ 1965 Neurosecretion and water balance in the male desert locust. *J. Zool. (Lond.) 147: 201-215* **brain NSC DH**

HIGHNAM KC HILL L MORDUE W 1966 The endocrine system and oocyte growth in *Schistocerca* in relation to starvation and frontal ganglionectomy. *J. Insect Physiol. 12: 977-994* **SGG retroc feed**

HILL KG 1983 The physiology of locust auditory receptors. I. Discrete depolarisations of receptor cells. *J. comp. Physiol. A 152: 475-482* **aud SO tymp**

HILL KG 1983 The physiology of locust auditory receptors. II. Membrane potentials associated with the response of the receptor cell. *J. comp. Physiol. A 152: 483-493* **aud SO tymp**

HILL RB 1963 The effect of acetylcholine on twitches of the locust leg. *Comp. Biochem. Physiol. 10: 203-208* **NM ACh TS mod musc**

HILL RB USHERWOOD PNR 1961 The action of 5-hydroxytryptamine and related compounds in neuromuscular transmission in the locust *Schistocerca gregaria. J. Physiol. (Lond.) 157: 393-401* **NM TS 5HT mod**

HINKLE M CAMHI JM 1972 Locust motoneurons: bursting activity correlated with axon diameter. *Science 175: 553-555* **MN flight resp**

HIRIPI L S-RÓZSA K 1980 Adenylate cyclase as a receptor for monoamines and a peptide transmitter proctolin in the insect brain. *In: Vizi ES Wollemann M (eds.), Aminergic and peptidergic receptors. Szeged Akad. Kiadó. Budapest: Pergamon Press. 57-69* **cAMP pep proct brain TS recept**

HIRIPI L S-RÓZSA K 1984 Octopamine - and dopamine - sensitive adenylate cylase in the brain of *Locusta migratoria* during its development. *Cell. mol. Neurobiol. 4: 199-206* **oct dopa cAMP brain devel enz**

HIRIPI L S-RÓZSA K MILLER TA 1979 The effect of proctolin on the adenylate and guanylate cyclases in the *Locusta* brain at various developmental stages. *Experientia 35: 1287-1288* **brain devel cAMP pep proct enz**

HIRIPI L S-RÓZSA KS 1973 Fluorometric determination of 5-hydroxytryptamine and catecholamines in the central nervous system and heart of *Locusta migratoria migratorioides. J. Insect Physiol. 19: 1481-1485* **TS 5HT jeart**

HIRIPI L S-RÓZSA KS 1987 Alterations of serotonin level in the central nervous system of *Locusta migratoria* during larval-adult transformation. *Acta biol. Hung. 38: 195-202* **TS 5HT devel**

HO RK BALL EE GOODMAN CS 1982 Muscle pioneers: large mesodermal cells that erect a scaffold for developing muscles and motoneurones in grasshopper embryos. *Nature 301: 66-69* **musc devel MN**

HO RK GOODMAN CS 1982 Peripheral pathways are pioneered by an array of central and peripheral neurones in grasshopper embryos. *Nature 297: 404-406* **devel**

HODGE C 1936 The anatomy and histology of the alimentary tract of the grasshopper *Melanoplus differentialis. J. Morphol. 59: 423-434* **anat gut**

HODGE C 1940 The anatomy and histology of the alimentary tract of *Radenotatum carinatum* var. *peninsulare* Rehn and Hebard (Orth., Acrid.). *J. Morphol. 66: 581-603* **anat gut**

HODGE C 1943 The internal anatomy of *Leptysma marginicollis* Serv. and of *Opshomala vitreipennis* Marsch. (Orth., Acrid.). *J. Morphol. 72: 87-123* **anat gut**

HODGKISS JP USHERWOOD PNR 1978 Transmitter release from normal and degenerating locust motor nerve terminals. *J. Physiol. (Lond.) 285: 113-128* **TS glu MN degen**

HOFFMANN JA GIRARDIE A 1969 Action du système endocrine et de l'ovaire sur l'hémogramme de *Locusta migratoria* L. (Orthoptère). Rôle de la pars intercerebralis. *C. R. Acad. Sci. Paris Ser. D 268: 1427-1429* **blood brain NSC**

HOFFMANN M PFLÜGER H-J 1990 Innervation and function of locust coxal rotator muscles. *In: Elsner N Roth G (eds.) Brain - Perception - Cognition: Proc. 18th Göttingen Neurobiol. Conf. Stuttgart, New York: Thieme Verlag. Abstract 38* **leg musc anat MN innerv**

HOLDEN JS SUTER C USHERWOOD PNR 1977 Isolation of neurone somata exhibiting pharmacological responses from the locust nervous system. *J. Physiol. (Lond.) 276: 4-5P* **TS glu cellCult**

HOLMAN GM NACHMAN RJ WRIGHT MS SCHOOFS L HAYES TK DE LOOF A 1990 Insect myotropic peptides: isolation, structural characterisation and biological activities. *In: Menn JJ Kelly TJ Masler EP (eds.), Insect neuropeptides: chemistry, biology and action. Washington DC: American Chemical Society. 40-50* **pep musc mod**

HOMBERG U 1990 The central complex in the brain of the locust: anatomical and physiological characterization. *In: Elsner N Roth G (eds.) Brain - Perception - Cognition: Proc. 18th Göttingen Neurobiol. Conf. Stuttgart, New York: Thieme Verlag. Abstract 318* **brain anat IN CB**

HOMBERG U 1991 Neurochemical anatomy of the central complex in the insect brain. *In: Elsner N Penzlin H (eds.), Synapse-transmission-modulation: Proc. 19th Göttingen Neurobiol. Conf. Stuttgart, New York: Thieme Verlag. Abstract 610* **anat brain CB TS pep**

HOMBERG U 1991 Neuroarchitecture of the central complex in the brain of the locusts *Schistocerca gregaria* and *S. americana* as revealed by serotonin immunocytochemistry. *J. comp. Neurol. 303: 1-10* **brain CB anat 5HT TS**

HOMBERG U 1991 Anatomical and functional characterization of the central complex in the brain of the locust *Schistocerca gregaria*. *Soc. Neurosci. Abstr. 17: 1244* **brain anat CB**

HOMBERG U BINKLE U 1989 Immunocytochemistry of neurotransmitters and peptides in the central complex of the locust brain. *In: Elsner N Singer W (eds.), Dynamics and plasticity in neuronal systems: Proc. 17th. Göttingen Neurobiology Conference. Stuttgart, New York: Thieme Verlag. Abstract 60* **brain CB anat TS pep**

HOMBERG U WURDEN S DIRCKSEN H RAO KR 1991 Comparative anatomy of pigment-dispersing hormone-immunoreactive neurons in the brain of orthopteroid insects. *Cell Tiss. Res. 266: 343-358* **brain pep anat CPDH rhythm**

HONEGGER H-W ALTMAN JS KIEN J MÜLLER-TAUTZ R POLLERBERG E 1984 A comparative study of neck muscle motor neurons in a cricket and a locust. *J. comp. Neurol. 230: 517-535* **neck thor MN anat gryll innerv musc**

HOPKINS TL YOUNG H 1990 Attraction of the grasshopper *Melanoplus sanguinipes* to host plant odors and volatile components. *Entomol. exp. appl. 56: 249-258* **feed chem behav**

HORN G ROWELL CHF 1968 Medium and long-term changes in the behaviour of visual neurones in the tritocerebrum of locusts. *J. exp. Biol. 49: 143-169* **DN habit vis**

HORRIDGE GA 1961 Pitch discrimination in locusts. *Proc. R. Soc. Lond. (Biol.) 155: 218-231* **aud SO IN tymp**

HORRIDGE GA 1962 Learning of leg position by headless insects. *Nature 193: 697-698* **learn leg thor**

HORRIDGE GA 1962 Learning of leg position by the ventral nerve cord in headless insects. *Proc. R. Soc. Lond. (Biol.) 157: 33-52* **learn thor leg**

HORRIDGE GA 1966 The retina of the locust. *In: Bernhard CG (ed.), The functional organization of the compound eye, London, Pergamon Press. 512-542* **ret anat**

HORRIDGE GA 1968 Affinity of neurones in regeneration. *Nature 219: 737-740* **regen anat**

HORRIDGE GA 1977 The compound eye of insects. *Sci. Am. 237: 108-121* **ret vis**

HORRIDGE GA 1978 The separation of visual axes in apposition compound eyes. *Phil. Trans. R. Soc. Lond. (Biol.) 285: 1-59* **ret SO vis**

HORRIDGE GA BARNARD PBT 1965 Movement of palisade in locust retinula cells when illuminated. *Quart. J. micr. Sci. 106: 131-135* **ret**

HORRIDGE GA MEINERTZHAGEN IA 1970 The exact neural projection of the visual fields upon the first and second ganglia of the insect eye. *Z. vergl. Physiol. 66: 369-379* **opt anat ret**

HORRIDGE GA SCHOLES JH SHAW S TUNSTALL J 1965 Extracellular recordings from single neurones in the optic lobe and brain of the locust. *In: Treherne JE Beament JWL (eds.), The physiology of the insect central nervous system. London: Academic Press. 165-202* **vis opt brain**

HORRIDGE GA TSUKAHARA Y 1978 The distribution of bumps in the tail of the locust photoreceptor afterpotential. *J. exp. Biol. 73: 1-14* **ret**

HORSEMAN G VIRANT-DOBERLET M LOHER W 1990 Neuroanatomy of the brain-retrocerebral complex in the grasshopper *Gomphocerus rufus. In: Elsner N Roth G (eds.) Brain - Perception - Cognition: Proc. 18th Göttingen Neurobiol. Conf. Stuttgart, New York: Thieme Verlag. Abstract 315* **retroc brain anat**

HORSMANN U 1981 Flugrelevante Afferenzen und ihre Verarbeitung bei der Wanderheuschrecke (*Locusta migratoria* L.). *Diplomarbeit, Universität Köln, Germany.* **flight propr**

HORSMANN U 1985 Der Einfluss propriozeptiver Windmessung auf den Flug der Wanderheuschrecke und die Bedeutung descendierender Neuronen der Tritocerebralkommisur. *Dissertation, Universität Köln, Germany.* **flight SO DN windh mech**

HORSMANN U HEINZEL HG WENDLER G 1983 The phasic influence of self-generated air current modulations on the locust flight motor. *J. comp. Physiol. A 150: 427-438* **flight windh SO mech DN propr**

HORSMANN U WENDLER G 1985 The role of a fast wing reflex in locust flight. *In: Gewecke M Wendler G (eds.), Insect Locomotion. Hamburg & Berlin: Paul Parey Verlag. 157-165* **flight SR propr**

HOU T GEWECKE M 1990 Movement sensitive visual neurons of the locust brain identified by video stimulation. *In: Elsner N Roth G (eds.) Brain - Perception - Cognition: Proc. 18th Göttingen Neurobiol. Conf. Stuttgart, New York: Thieme Verlag. Abstract 210* **vis IN tech**

HOWARD J 1981 Temporal resolving power of the photoreceptors of *Locusta migratoria. J. comp. Physiol. A 144: 61-66* **ret vis**

HOWARD J DUBS A PAYNE R 1984 The dynamics of phototransduction in insects. A comparative study. *J. comp. Physiol. A 154: 707-718* **ret**

HOWSE PE 1967 Mechanism of the insect ear. *Nature 213: 367-369* **tymp subg aud SO**

HOWSE PE LEWIS DB PYE JD 1971 Adequate stimulus of the insect tympanic organ. *Experientia 27: 598* **SO tymp aud**

HOY R NOLEN T BRODFUEHRER P 1989 The neuroethology of acoustic startle and escape in flying insects. *J. exp. Biol. 146: 209-228* **aud flight gryll dict lep IN thor**

HOYLE G 1952 High blood potassium in insects in relation to nerve conduction. *Nature 169: 281-282* **blood**

HOYLE G 1953 Potassium ions and insect nerve/muscle. *J. exp. Biol. 30: 121-135* **NM saline blood**

HOYLE G 1953 Intracellular recording of "slow" and "fast" nerve fibres in locusts. *J. Physiol. (Lond.) 121: 32P* **musc**

HOYLE G 1953 "Slow" and "fast" nerve fibres in locusts. *Nature 172: 165-167* **musc innerv**

HOYLE G 1954 Changes in the blood concentration of the african migratory locust (*Locusta migratoria migratorioides* R. & F.) during food deprivation, and the effect on neuromuscular activity. *J. exp. Biol. 31: 260-270* **NM saline blood**

HOYLE G 1954 Functioning of the insect ocellar nerve. *J. exp. Biol. 32: 397-407* **oc brain**

HOYLE G 1955 The effects of some common cations on neuromuscular transmission in insects. *J. Physiol. (Lond.) 127: 90-103* **NM blood**

HOYLE G 1955 The anatomy and innervation of locust skeletal muscle. *Proc. R. Soc. Lond. (Biol.) 143: 281-292* **musc anat innerv**

HOYLE G 1955 Neuromuscular mechanism of a locust skeletal muscle. *Proc. R. Soc. Lond. (Biol.) 143: 343-367* **musc NM**

HOYLE G 1956 Sodium and potassium changes occurring in the haemolymph of insects at the time of moulting and their physiological consequences. *Nature 178: 1236-1237* **NM saline moult blood**

HOYLE G 1957 Nervous control of insect muscles. *In: Scheer BT (ed.), Recent advances in invertebrate physiology. A symposium. Eugene: University of Oregon Publications. 73-98* **musc MN NM**

HOYLE G 1958 The leap of the grasshopper. *Sci. Am. 198: 30-35* **jump**

HOYLE G 1959 The neuromuscular mechanism of an insect spiracular muscle. *J. Insect Physiol. 3: 378-394* **NM musc resp spir**

HOYLE G 1960 The action of carbon dioxide on an insect spiracular muscle. *J. Insect Physiol. 4: 63-79* **resp spir musc**

HOYLE G 1961 Functional contracture in a spiracular muscle. *J. Insect Physiol. 7: 305-314* **musc resp spir**

HOYLE G 1964 Exploration of neuronal mechanisms underlying behavior in insects. *In: Reiss RF (ed.), Neural theory and modeling. Stanford, California: Stanford University Press. 346-376* **IN MN**

HOYLE G 1965 Neurophysiological studies on "learning" in headless insects. *In: Treherne JE Beament JWL (eds.), The physiology of the insect central nervous system. New York, London: Academic Press. 203-232* **learn MN leg**

HOYLE G 1966 An isolated insect ganglion-nerve-muscle preparation. *J. exp. Biol. 44: 413-428* **NM leg musc tech**

HOYLE G 1966 Functioning of the inhibitory conditioning axon innervating insect muscles. *J. exp. Biol. 44: 429-453* **musc inhib leg NM**

HOYLE G 1974 Neural control of skeletal muscle. *In: Rockstein M (ed.), The physiology of Insecta. Vol. IV (2nd Edition). 175-236* **musc innerv MN rev**

HOYLE G 1974 A function for neurons (DUM) neurosecretory on skeletal muscle of insects. *J. exp. Zool. 189: 401-406* **DUM musc**

HOYLE G 1975 Evidence that insect dorsal unpaired medican *(sic)* (DUM) neurons are octopaminergic. *J. exp. Zool. 193: 425-431* **DUM TS oct**

HOYLE G 1975 Identified neurons and the future of neuroethology. *J. exp. Zool. 194: 51-73* **anat behav**

HOYLE G 1975 Modulation of intrinsic rhythmicity of insect muscle by (octopaminergic) DUM neurons. *Soc. Neurosci. Abstracts 5: 567* **DUM rhythm musc oct leg mod**

HOYLE G 1978 Intrinsic rhythm and basic tonus in insect skeletal muscle. *J. exp. Biol. 73: 173-203* **musc innerv leg rhythm NM**

HOYLE G 1978 Distribution of nerve and muscle fibre types in locust jumping muscle. *J. exp. Biol. 73: 205-233* **musc DUM leg innerv**

HOYLE G 1978 The dorsal unpaired median neurons of the locust metathoracic ganglion. *J. Neurobiol. 9: 43-57* **DUM thor**

HOYLE G 1980 Learning, using natural reinforcements, in insect preparations that permit cellular neuronal analysis. *J. Neurobiol. 11: 323-354* **learn MN leg**

HOYLE G 1982 Pacemaker change in a learning paradigm. *In: Carpenter D (ed.), Cellular Pacemakers, Vol. 2. New York: J. Wiley & Sons Inc. 3-25* **learn MN leg**

HOYLE G 1985 Generation of behaviour: the orchestration hypothesis. *In: Barnes WJP Gladden MH (eds.), Feedback and motor control in invertebrates and vertebrates. London, Sydney, Dover, New Hampshire: Croom Helm Ltd. 57-75* **mod oct**

HOYLE G 1985 Generation of motor activity and control of behaviour: The roles of neuromodulator octopamine, and the orchestration hypothesis. *In: Kerkut GA Gilbert LI (eds.), Comprehensive insect physiology, biochemistry and pharmacology, Vol. 5. London: Pergamon Press. 607-621* **mod oct**

HOYLE G 1985 Glia cells of an insect ganglion. *J. comp. Neurol. 246: 85-103* **glia anat thor**

HOYLE G BARKER D 1975 Synthesis of octopamine by insect dorsal median unpaired neurons. *J. exp. Zool. 193: 433-439* **DUM TS oct**

HOYLE G BURROWS M 1970 Intracellular studies on identified neurons of insects. *Fedn. Proc. 29: 589* **IN thor**

HOYLE G BURROWS M 1973 Neural mechanisms underlying behavior in the locust *Schistocerca gregaria.* I. Physiology of identified motoneurons in the metathoracic ganglion. *J. Neurobiol. 4: 3-41* **MN leg**

HOYLE G BURROWS M 1973 Neural mechanisms underlying behavior in the locust *Schistocerca gregaria.* II. Integrative activity in metathoracic neurons. *J. Neurobiol. 4: 43-67* **IN thor**

HOYLE G COLQUHOUN W WILLIAMS M 1980 Fine structure of an octopaminergic neuron and its terminals. *J. Neurobiol. 11: 103-126* **TS oct DUM fineStruc**

HOYLE G DAGAN D 1978 Physiological characteristics and reflex activation of DUM (octopaminergic) neurons of locust metathoracic ganglion. *J. Neurobiol. 9: 59-79* **DUM oct TS**

HOYLE G DAGAN D MOBERLY B COLQUHOUN W 1974 Dorsal unpaired median nerve insect neurons make neurosecretory endings on skeletal muscle. *J. exp. Zool. 187: 159-165* **DUM NM NSC leg**

HOYLE G O'SHEA M 1974 Intrinsic rhythmic contractions in insect skeletal muscle. *J. exp. Zool. 189: 407-412* **rhythm musc leg**

HOYLE G SIDIE JM 1976 Temporal patterning in spike trains from an insect motorneuron. *Soc. Neurosci. Abstr. 6: 334* **MN**

HOYLE G WILLIAMS M PHILLIPS C 1986 Functional morphology of insect neuronal cell-surface/glial contacts: the trophospongium. *J. comp. Neurol. 246: 113-128* **glia fineStruc**

HSÜ F LIU W-T SHEN L-M 1952 [The abdominal tympanal organs of some Acrididae] (in Chinese, with engl. summary). *Acta entomol. Sin. 10: 53-59* **tymp SO anat**

HUBBARD SJ 1959 Femoral mechanoreceptors in the locust. *J. Physiol. (Lond.) 147: 8-10P* **FCO SO propr leg**

HUBER F 1955 Über die Funktion der Pilzkörper (corpora pedunculata) beim Gesang der Keulenfeldheuschrecke *Gomphocerus rufus* L. (Acrididae). *Naturwiss. 42: 566-567* **brainStim strid**

HUBER F 1959 Auslösung von Bewegungsmustern durch elektrische Reizung des Oberschlundganglions bei Orthopteren (Saltatoria, Gryllidae, Acrididae). *Verh. dt. zool. Ges. 1959: 248-269* **strid brainStim gryll**

HUBER F 1964 The role of the CNS in Orthoptera during the coordination and control of stridulation. *In: Busnel RG (ed.): Acoustic behaviour of animals. Amsterdam: Elsevier Publishing Co. 440-488* **strid brainStim**

HUDDART H 1971 Ultrastructure of a peripheral nerve of *Locusta migratoria migratorioides* R. & F. (Orthoptera: Locustidae). *Int. J. Insect Morphol. Embryol. 1: 29-41* **fineStruc leg sheath glia trachea**

HUDDART H 1985 Visceral muscle. *In: Kerkut GA Gilbert LI (eds.), Comprehensive insect physiology, biochemistry and pharmacology, Vol. 11. London: Pergamon Press. 131-194* **musc gut genit**

HUDDART H 1986 Influence of glutamate, dopamine and 5-hydroxytryptamine on calcium efflux from locust foregut and hindgut visceral muscle. Dependence upon extracellular calcium and calcium channel entry. *Comp. Biochem. Physiol. C 84: 401-408* **musc gut innerv TS glu dopa 5HT Ca2 channel**

HUDDART H ABRAM GR 1969 Modification of excitation-contraction coupling in locust skeletal muscle induced by caffeine. *J. exp. Zool. 171: 49-58* **musc Ca2**

HUDDART H OATES K 1970 Ultrastructure of stick insect and locust skeletal muscle in relation to excitation-contraction coupling. *J. Insect Physiol. 16: 1467-1485* **musc fineStruc**

HUDDART H OLDFIELD AC 1982 Spontaneous activity of foregut and hindgut visceral muscle of the locust, *Locusta migratoria*. II. The effect of biogenic amines. *Comp. Biochem. Physiol. C 73: 303-311* **gut innerv TS rhythm 5HT oct**

HUDDIE PL RAMSEY RL SANSOM MSP USHERWOOD PNR 1986 Kinetics of cysteine sulphinate gated ion channels of locust (*Schistocerca gregaria*) muscle. *Neurosci. Lett. Suppl. 24: 33 (Abstract)* **channel musc**

HUDDIE PL RAMSEY RL USHERWOOD PNR 1986 Single potassium channels of adult locust (*Schistocerca gregaria*) muscle recorded using the giga-ohm seal patch-clamp technique. *J. Physiol. (Lond.) 378: 60P* **channel musc**

HUDDIE PL USHERWOOD PNR 1989 Ryanodine modifies the ion selectivity of potassium channels of locust muscle. *In: Lauger P Packer L Vasilescu V (eds.): Water and ions in biological systems. Basel, Birkhäuser Verlag, 151-161* **channel musc**

HUGHES GM 1952 Abdominal mechanoreceptors in *Dytiscus* and *Locusta. Nature 170: 531* **propr SO abd**

HUGHES TD 1980 The imaginal ecdysis of the desert locust, *Schistocerca gregaria.* I. A description of the behaviour. *Physiol. Entomol. 5: 47-54* **moult behav**

HUGHES TD 1980 The imaginal ecdysis of the desert locust, *Schistocerca gregaria.* II. Motor activity and emergence behaviour. *Physiol. Entomol. 5: 55-71* **moult rhythm**

HUGHES TD 1980 The imaginal ecdysis of the desert locust, *Schistocerca gregaria.* III. Motor activity underlying the expansional and post-expansional behaviour. *Physiol. Entomol. 5: 141-152* **moult**

HUGHES TD 1980 The imaginal ecdysis of the desert locust, *Schistocerca gregaria.* IV. The role of the gut. *Physiol. Entomol. 5: 153-164* **moult gut**

HUSTERT R 1974 Morphologie und Atmungsbewegungen des 5. Abdominalsegmentes von *Locusta migratoria migratorioides. Zool. Jb. (Physiol.) 78: 157-174* **resp anat abd musc**

HUSTERT R 1975 Neuromuscular coordination and proprioceptive control of rhythmical abdominal ventilation in intact *Locusta migratoria migratorioides. J. comp. Physiol. A 97: 159-179* **SO resp propr**

HUSTERT R 1978 Segmental and interganglionic projections from primary fibres of insect mechanoreceptors. *Cell Tiss. Res. 194: 337-351* **SO mech anat thor leg abd**

HUSTERT R 1981 Distribution and specific central projections of mechanoreceptors in the thorax and proximal leg joints of locusts. III. The external mechanoreceptors: the campaniform sensilla. *Cell Tiss. Res. 216: 97-111* **SO anat mech thor leg**

HUSTERT R 1982 The proprioceptive function of a complex chordotonal organ associated with the mesothoracic coxa in locusts. *J. comp. Physiol. A 147: 389-399* **SO thor propr MN FCO**

HUSTERT R 1983 Proprioceptor responses and convergence of proprioceptive influence on motoneurones in the mesothoracic thoraco-coxal joint of locusts. *J. comp. Physiol. A 150: 77-86* **propr leg MN thor anat**

HUSTERT R 1985 The contribution of proprioceptors to the control of motor patterns of legs in orthopterous insects - the locust example. *In: Gewecke M Wendler G (eds.), Insect locomotion. Hamburg & Berlin: Paul Parey Verlag. 59-67* **propr leg walk rev**

HUSTERT R 1988 Normal and regenerated tarsal mechanoreceptors control walking in locusts. *In: Elsner N Barth FG (eds.) Sense organs: interfaces* *between environment and behaviour: Proc.16th. Göttingen Neurobiol. Conf. Stuttgart, New York: Thieme Verlag. Abstract 374* **SO leg walk regen mech**

HUSTERT R 1989 Mesothoracic interneurons may coordinate the legs of locusts. *In: Erber J Menzel R Pflüger H-J Todt D (eds.), Neural mechanisms of behavior: Proc. 2nd Internat. Congr. Neuroethology. Stuttgart & New York: Thieme Verlag. Abstract 17* **IN thor propr walk leg anat**

HWANG G-H LONG C-C 1962 [The change of body temperature of the oriental migratory locust (*Locusta migratoria manilensis* Meyen) during flight.] (in Chinese, with English summary).. *Acta entomol. Sin. 11: 419-421* **temp flight**

INGLIS M OLDFIELD BP 1988 Tonotopic organization of the auditory organ of the locust *Valanga irregularis* (Walker). *J. comp. Physiol. A 164: 49-53* **aud SO tymp**

IRVING SN WILSON RG OSBORNE MP 1979 Studies on L-glutamate in insect haemolymph. III. Aminoacid analyses of the haemolymph of various arthropods. *Physiol. Entomol. 4: 231-240* **blood glu**

ISAAC RE 1987 Proctolin degradation by membrane peptidases from nervous tissues of the desert locust *Schistocerca gregaria. Biochem. J. 245: 365-370* **pep TS proct enz**

ISAAC RE 1988 Neuropeptide-degrading endopeptidase activity of locust (*Schistocerca gregaria*) synaptic membranes. *Biochem. J. 255: 843-847* **pep TS enz**

IVANOVA TS 1952 [Anatomy of the peripheral nervous system of the abdominal segments of the asiatic locust (*Locusta migratoria* L.).] (in Russian) *Entomol. Obozr. 32: 148-159* **anat abd**

IVANOVA TS 1956 [Innervation of skeleton muscles by the system of paired ventral nerves in *Locusta migratoria* L. (Orthoptera, Acrididae)]. (in Russian) *Entomol. Obozr. 35: 782-788* **anat musc innerv**

IVLEV SV LIPSKAJA EF MANDELSHTAM YE DEIKO LI 1989 Effects of dineopentyl and dipinacoline ethers of glutamic acid on neuromuscular transmission in *Locusta migratoria. Gen. Physiol. Biophys. 8: 163-170* **glu NM TS tox**

JABBAR A STRANG RHC 1980 A method for the qualitative study of amines and amino-acids of the insect nervous system. *In: Sherwood M (ed.): Insect neurobiology and pesticide action (Neurotox 79). London: Society of Chemical Industry Press. 261-266* **metab tech TS**

JABBAR A STRANG RHC 1981 Synthesis of alanine in the nervous tissue of the locust (*Schistocerca americana gregaria*). *Proc. 8th Int. Congr. Neurochem.: 187* **metab**

JABBAR A STRANG RHC 1985 The effects of chemical and physical stress on the concentrations of amino compounds in the haemolymph and nervous system of locusts and cockroaches. *J. Insect Physiol. 31: 359-370* **blood dict**

JABBAR A STRANG RHC 1987 Anaerobic aminoacid metabolism in the locust nervous system. *Comp. Biochem. Physiol. 88B : 65-69* **metab**

JABBAR A STRANG RHC 1987 Aerobic aminoacid metabolism in the locust nervous system. *Insect Biochem. 17: 207-211* **metab**

JAHN TL CRECITELLI F 1938 The electrical response of the grasshopper eye under conditions of light and dark adaptation. *J. cell. comp. Physiol. 12: 39-55* **ret**

JAHN TL CRECITELLI F 1940 Diurnal changes in the electrical response of the compound eye. *Biol. Bull. Woods Hole 78: 42-52* **ret**

JAHN TL WULFF VJ 1942 The effect of temperature on the retinal action potential. *Anat. Rec. 84: 535* **ret temp**

JAHN TL WULFF VJ 1942 Allocation of electrical responses from the compound eye of grasshoppers. *J. gen. Physiol. 26: 75-88* **ret**

JAHROMI SS BLOOM JW 1979 Structural changes in locust leg muscle fibres in response to enotomy and joint mobilization. *J. Insect Physiol. 25: 767-780* **musc**

JAILLET I 1969 Enregistrement de l'activité de trois neurones dans les sensilla basiconica cercales de *Locusta migratoria* L. (Orthoptera, Acrididae). *C. R. Seanc. Soc. Biol., Paris 163: 92-96* **SO cerc abd**

JAILLET I 1969 Contribution à l'étude de l'équipement sensoriel de l'extrémité abdominale de la femelle du criquet migrateur *Locusta migratoria* L. (Orthoptera, Acrididae). *Rev. comp. Anim. 3: 33-77* **anat SO abd genit**

JAN L-Y JAN Y-N 1982 Antibodies to horseradish peroxidase as specific neuronal markers in *Drosophila* and grasshopper embryos. *Proc. nat. Acad. Sci. USA 79: 2700-2704* **tech devel**

JANNONE G 1939 Studio morfologico, anatomico e istologico del *Dociostaurus maroccanus* Thunberg, nelle sue fasi *transiens congregans, gregaria* e *solitaria*. *Boll. R. Lab. entomol. agr. Portici 4: 1-443* **anat phase**

JANSEN WF DIEDEREN JHB DORLAND M LANGERMANS J MEESEN BPM MINK K VULLNGS HGB 1989 Ultrastructural enzyme-cytochemical study of the intrinsic glandular cells in the corpus cardiacum of *Locusta migratoria*: relation to the secretory and endocytotic pathways, and to the lysosomal system. *Cell Tiss. Res. 255: 167-178* **enz fineStruc retroc CC**

JAWLOWSKI H 1954 Über die Struktur des Gehirnes bei Saltatoria. *Ann. Univ. M. Curie-Sklodowska C 8: 403-434* **anat brain**

JAY DG KESHISHIAN H 1990 Laser inactivation of fasciclin I disrupts axon adhesion of grasshopper pioneer neurons. *Nature 348: 548* **devel**

JENSEN M 1956 Biology and physics of locust flight. III. The aerodynamics of flight. *Phil. Trans. R. Soc. Lond. (Biol.) 239: 511-532* **flight aero**

JERAM S COKL A 1991 Comparative physiology and anatomy of antennal mechanoreceptors in land bugs and locusts. *In: Elsner N Penzlin H (eds.), Synapse-transmission-modulation: Proc. 19th Göttingen Neurobiol. Conf. Stuttgart, New York: Thieme Verlag. Abstract 50* **ant SO mech**

JOHNSON AG ROBERTSON RM LYWOOD DW 1991 Steering torques generated by collision avoidance manoevres during flight in the locust. *Soc. Neurosci. Abstr. 17: 1244* **behav flight OM**

JOLY L JOLY P PORTE A GIRARDIE A 1968 Etude physiologique et ultrastructurale des corpora allata de *Locusta migratoria* L. (Orthoptère) en phase grégaire. *Arch. Zool. exp. gén. 109: 703-728* **retroc CAll anat fineStruc**

JOLY L JOLY P PORTE A GIRARDIE A 1969 Analyse ultrastructurale de l'activité des corpora allata de *Locusta migratoria* et ses conséquences sur la structure que l'on doit attribuer au mécanisme humoral contrôlant la métamorphose. *Archs. Zool. exp. gén. 110: 617-628* **retroc CAll fineStruc**

JOLY P 1951 Rôle de l'organe de Brunner chez *Locusta migratoria* (L.). *C. R. Seanc. Soc. Biol., Paris 145: 1354-1357* **SO leg mech brun**

JOLY P GIRARDIE A 1967 Caractères ultrastructuraux des corpora allata actifs et inactifs chez *Locusta migratoria. C. R. Acad. Sci. Paris Ser. D 265: 1633-1635* **retroc CAll anat fineStruc**

JORGENSEN WK RICE MJ 1982 Superextension and supercontraction in locust ovipositor muscles. *J. Insect Physiol. 29: 437-448* **musc genit ovip**

JÖRSCHKE H 1914 Die Facettenaugen der Orthopteren und Termiten. *Z. wiss. Zool. 111:153-280* **vis SO mech ret**

KAFKA WA 1970 Molekulare Wechselwirkungen bei der Erregung einzelner Riechzellen. *Z. vergl. Physiol. 70: 105-143* **ant SO**

KAFKA WA OHLOFF G SCHNEIDER D VARESCHI E 1973 Olfactory discrimination of two enantiomers of 4-methyl-hexanoic acid by the migratory locust and the honeybee. *J. comp. Physiol. 87: 277-284* **ant SO chem**

KALLADUR VL DOWNER RGH GEORGE JC THOMPSON JE 1982 Effect of environmental temperature on the phase properties and lipid composition of flight muscle mitochondria of *Schistocerca gregaria. Insect Biochem. 12: 115-121* **musc flight**

KALMRING K 1971 Akustische Neuronen im Unterschlundganglion der Wanderheuschrecke *Locusta migratoria. Z. vergl. Physiol. 72: 95-110* **aud IN suboes**

KALMRING K 1975 The afferent auditory pathway in the ventral cord of *Locusta migratoria* (Acrididae). I. Synaptic connectivity and information processing among the auditory neurons of the ventral cord. *J. comp. Physiol. A 104: 103-141* **aud IN SO tymp thor**

KALMRING K 1975 The afferent auditory pathway in the ventral cord of *Locusta migratoria* (Acrididae). II. Responses of the auditory ventral cord neurons to natural sounds. *J. comp. Physiol. A 104: 143-159* **aud IN habit strid thor**

KALMRING K 1983 Convergence of auditory and vibratory senses at the neuronal level in the ventral nerve cord in grasshoppers: its probable importance for behaviour in the habitat. *In: Horn E (ed.), Multimodal convergence in sensory systems. Stuttgart, New York: Gustav Fischer Verlag. 129-141* **aud IN vibr**

KALMRING K KAISER W OTTO C KÜHNE R 1985 Coprocessing of vibratory and auditory information in the CNS of different tettigoniids and locusts. *In: Kalmring K Elsner N (eds.) Acoustic and vibrational communication in insects. Hamburg & New York: Paul Parey Verlag. 193-202* **aud vibr IN**

KALMRING K KÜHNE R MOYSICH F 1978 The auditory pathway in the central cord of the migratory locust (*Locusta migratoria*): response transmission in the axons. *J. comp. Physiol. A 126: 25-33* **aud IN**

KALMRING K KÜHNE R MOYSICH F 1978 The coding of sound signals in the ventral cord auditory system of the migratory locust (*Locusta migratoria*). *J. comp. Physiol. A 128: 213-226* **aud IN thor**

KALMRING K RHEINLÄNDER J REHBEIN H 1972 Akustische Neuronen im Bauchmark der Wanderheuschrecke *Locusta migratoria. Z. vergl. Physiol. 76: 314-332* **aud IN thor**

KALMRING K RHEINLÄNDER J RÖMER H 1972 Akustische Neuronen im Bauchmark von *Locusta migratoria.* Der Einfluss der Schallrichtung auf die Antwortmuster. *J. comp. Physiol. 80: 325-352* **aud IN thor**

KALMRING K RÖMER H REHBEIN H 1974 Connections of acoustic neurons within the CNS of grasshoppers. *Naturwiss. 61: 454* **aud IN MN**

KALOGIANNI E 1989 Locust oviducal motor neurons. *In: Erber J Menzel R Pflüger H-J Todt D (eds.), Neural mechanisms of behavior: Proc. 2nd Internat. Congr. Neuroethology. Stuttgart & New York: Thieme Verlag. Abstract 48* **MN anat genit**

KAMEDA K 1964 Electrical activity of optic ganglion cells in locust (Orthoptera). *Bull. Tokyo med. dent. Univ. 11: 281-295* **opt brain vis IN**

KAMMER AE 1985 Flying. *In: Kerkut GA Gilbert LI (eds.), Comprehensive insect physiology, biochemistry and pharmacology Vol. 8. London: Pergamon Press. 491-552* **flight behav**

KÄMPER G RIEDE K 1989 Comparison of the auditory system of singing and silent grasshoppers. *In: Erber J Menzel R Pflüger H-J Todt D (eds.), Neural mechanisms of behavior: Proc. 2nd Internat. Congr. Neuroethology. Stuttgart & New York: Thieme Verlag. Abstract 64* **aud SO tymp anat**

KARANDIKAR KR 1939 External structures of the desert locust (*Schistocerca gregaria* Forsk.). *J. Univ. Bombay B 7: 56 pp.* **anat**

KARANDIKAR KR 1942 External structures of the desert locust (*Schistocerca gregaria* Forsk.). *J. Univ. Bombay B 11: 29 pp.* **anat**

KARELIN YA 1978 [Scanning electron microscopy of the head receptive hairs responsible for the maintenance of flight in *Locusta migratoria.*] (in Russian). *In: [Morphological Principles of Evolutionary Physiology.] Leningrad: Nauka. 30-34* **windh SO mech fineStruc**

KARELIN YA 1980 [Structural and functional organization of the afferent system involved in the function of flight mechanism in insects.] (in Russian). *Cand. Sci. Thesis, Leningrad, Sechenov Inst.* **windh SO mech flight**

KARELIN YA PLOTNIKOVA SI 1983 [Sensory-motor region of the deuterocerebrum in the locust *Locusta migratoria* L.]. (In Russian) *J. evol. Biochem. Physiol. 19 (3): 312-316* **brain anat ant**

KARLSTROM RO BASTIANI MJ 1990 Further characterization and cloning of a surface antigen expressed in a striped pattern during segmentation and later specific to the nervous system in grasshopper embryos. *Soc. Neurosci. Abstr. 16: 303* **devel**

KARST H FOKKENS RH DE HAAN N HEUVER G HUE B KRUK C NIBBERING NMM PIEK T SPANJER W TONG YC VAN DER VLIET W 1991 Beta-philanthotoxin, a novel glutamatergic antagonist for insect neuromuscular transmission. *Comp. Biochem. Physiol. C 97: 317-328* **tox glu TS NM ACh**

KARST H HUE B PIEK T 1990 1-naphthylacetylspermine blocks glutamatergic and nicotinic synaptic transmission like philanthotoxin. *Neurosci. Res. Comm. 7: 61-68* **tox glu TS NM ACh**

KARST H PIEK T 1991 Structure-activity relationship of philanthotoxins. II. Effects on the glutamate-gated ion channels of the locust muscle fibre membrane. *Comp. Biochem. Physiol. C 98: 479-489* **tox glu musc**

KARST H PIEK T VAN MARLE J 1990 Dissociation of the presynaptic and postsynaptic inhibiting properties of the glutamatergic transmitter antagonist, philanthotoxin (PTX-4.3.3). *Neurosci. Res. Comm. 7: 69-74* **tox TS glu NM**

KARST H PIEK T VAN MARLE J LIND A VAN WEEREN-KRAMER J 1991 Structure-activity relationship of philanthotoxins. I. Presynaptic and postsynaptic inhibition of the locust neuromuscular transmission. *Comp. Biochem. Physiol. C 98: 471-478* **tox TS glu NM**

KAUFMANN L BENSON JA 1991 Characterisation of a locust neuronal octopamine response. *Soc. Neurosci. Abstr. 17: 277* **cellCult oct tox recept**

KATSUKI Y SUGA N 1960 Neural mechanisms of hearing in insects. *J. exp. Biol. 37: 279-290* **aud SO tymp**

KATZ SL 1991 Ontogenetic scaling of locust jump performance. *Am. Zool. 31: 141A* **behav jump devel**

KAWAMURA K 1960 Behavior of spindle body in neuroblasts of the grasshopper, *Chortophaga viridifasciata* (De. Geer). (Abstract). *Ass. Southeastern Biol. Bull. 7: 32* **devel**

KAWAMURA K 1960 Studies on cytokinesis in neuroblasts of the grasshopper *Chortophaga viridifasciata* (De Geer). I. Formation and behavior of the mitotic apparatus. *Exptl. Cell Res. 21: 1-8* **devel**

KAWAMURA K 1960 Studies on cytokinesis in neuroblasts of the grasshopper *Chortophaga viridifasciata* (De Geer). II. The role of the mitotic apparatus in cytokinesis. *Exptl. Cell Res. 21: 9-18* **devel**

KAWAMURA K 1977 Microdissection studies on the dividing neuroblast of the grasshopper, with special reference to the mechanism of unequal cytokinesis. *Exptl. Cell Res. 106: 127-137* **devel**

KAWAMURA K 1988 The contraction wave in the cortex of dividing neuroblasts of the grasshopper. *Zool. Sci. 5: 677-684* **devel**

KAWAMURA K CARLSON JG 1962 Studies on cytokinesis in neuroblasts of the grasshopper *Chortophaga viridifasciata* (De Geer). III. Factors determing the location of the cleavage furrow. *Exptl. Cell Res. 26: 411-423* **devel**

KAWAMURA K YAMASHIKI N 1990 Factors inducing unequal cytokinesis in grasshopper. *Ann. N. Y. Acad. Sci. 582: 304-306* **devel**

KENDALL M SEDDON A 1975 The effect of previous access to water on the responses of locusts to wet surfaces. *Acrida 4: 1-8* **SO ant leg hygr**

KENDALL MD 1970 The anatomy of the tarsi of *Schistocerca gregaria* Forskål. *Z. Zellforsch. mikr. Anat. 109: 112-137* **anat leg SO CO**

KENDALL MD 1972 Glandular epidermis on the tarsi of the desert locust *Schistocerca gregaria* Forskål. *Acrida 1: 121-149* **anat pher leg**

KENDIG JJ 1968 Motor neurone coupling in locust flight. *J. exp. Biol. 48: 389-404* **flight MN**

KERRY CJ KITS KS RAMSEY RL SANSOM MSP USHERWOOD PNR 1987 Single channel kinetics of a glutamate receptor. *Biophys. J. 137-144* **musc NM glu channel TS recept**

KERRY CJ RAMSEY RL SANSOM MSP USHERWOOD PNR 1988 Glutamate receptor channel kinetics - the effect of glutamate concentration. *Biophys. J. 53: 39-52* **TS glu musc channel recept**

KERRY CJ RAMSEY RL SANSOM MSP USHERWOOD PNR 1988 Single channel studies of non-competitive antagonism of a quisqualate-sensitive glutamate receptor by argiotoxin 636 - a fraction isolated from orb-web spider venom. *Brain Res. 459: 312-327* **TS glu musc tox recept**

KERRY CJ RAMSEY RL SANSOM MSP USHERWOOD PNR WASHIO H 1985 The effect of tubocurarine on single glutamate-activated channels in locust muscle. *J. Physiol. (Lond.) 365: 82P* **TS glu musc channel tox**

KERRY CJ RAMSEY RL SANSOM MSP USHERWOOD NPR WASHIO H 1987 Single-channel studies of the action of (+)-tubocurarine on locust muscle glutamate receptors. *J. exp. Biol. 127: 121-134* **TS glu musc tox recept channel**

KERRY CJ SANSOM MSP USHERWOOD PNR 1988 Molecular pharmacology of single glutamate receptor-channels. *Neurosci. Lett. Suppl. 32, S3 (Abstract)* **TS glu channel recept NM**

KERRY CJ USHERWOOD PNR 1987 Interaction of spider toxins with the locust glutamate receptor. *Neurosci. 22: 248 P* **TS glu channel recept NM tox**

KESHISHIAN H 1980 Origin and morphogenesis of pioneer neurons in the grasshopper metathoracic leg. *Dev. Biol. 80: 388-397* **devel leg**

KESHISHIAN H 1987 The cellular mechanisms that shape neuronal diversity and form. Insights from the embryonic grasshopper. *In: Gupta AP (ed.), Arthropod brain: its evolution, development, structure and function. New York: John Wiley & Sons Inc. 43-62* **devel leg**

KESHISHIAN H BENTLEY D 1983 Embryogenesis of peripheral nerve pathways in grasshoppers legs. I. The initial nerve pathway to the CNS. *Dev. Biol. 96: 89-102* **devel leg**

KESHISHIAN H BENTLEY D 1983 Embryogenesis of peripheral nerve pathways in grasshoppers legs. II. The major nerve routes. *Dev. Biol. 96: 103-115* **devel leg**

KESHISHIAN H BENTLEY D 1983 Embryogenesis of peripheral nerve pathways in grasshoppers legs. III. Development without pioneer neurons. *Dev. Biol. 96: 116-124* **devel leg**

KESHISHIAN H O'SHEA M 1984 Development, acquisition and expression of a peptide neurotransmitter in a model neuromuscular system of the grasshopper embryo. *Soc. Neurosci. Abstr. 10: 513* **pep proct devel NM**

KESHISHIAN H O'SHEA M 1985 Aquisition & expression of a peptidergic phenotype in the grasshopper embryo. *J. Neurosci. 5: 1005-1015* **proct pep**

KESHISHIAN H O'SHEA M 1985 The distribution of a peptide neurotransmitter in the grasshopper postembryonic central nervous system. *J. Neurosci. 5: 992-1004* **proct pep**

KEUPER A 1984 Die Untersuchung von multimodalen Konvergenzen und sensomotorischen Übergängen im ZNS von *Locusta migratoria*: Mehrkanalelektroden-Ableitung und Stromquellendichtemethode. *Dissertation, Universität Marburg, Germany.* **tech**

KEUPER A GREUEL J BRECKOW J 1985 The senso-motor-transition of vibro-acoustic signals in the CNS of *Locusta migratoria. In: Kalmring K Elsner N (eds.) Acoustic and vibrational communication in insects. Hamburg & New York: Paul Parey Verlag. 203-214* **aud vibr IN**

KHAN HR AZIZ SA 1974 Preliminary observations on antennal sense organs of *Eyprepocnemis alacris* Serville (Orthoptera: Acrididae). *Indian J. Entomol. 34: 353-355* **ant SO anat**

KHAN MA 1988 Brain-controlled synthesis of juvenile hormone in adult insects. *Entomol. exp. appl. 46: 3-18* **JH retroc CAll brain**

KIEN J 1974 Sensory integration in the locust optomotor system. I. Behavioural analysis. *Vision Res. 14: 1245-1254* **vis DN opt OM**

KIEN J 1974 Sensory integration in the locust optomotor system. II. Direction selective neurons in the circumoesophageal connectives and the optic lobe. *Vision Res. 14: 1255-1268* **vis DN opt OM**

KIEN J 1975 Motion detectors in locusts and grasshoppers. *In: Horridge GA (ed.), The compound eye and vision of insects. Oxford: Clarendon Press. 410-422* **vis opt IN OM**

KIEN J 1975 Neuronal mechanisms subserving directional selectivity in the locust optomotor system. *J. comp. Physiol. A 102: 337-355* **vis opt IN OM**

KIEN J 1976 A preliminary report of cobalt sulphide staining of locust visual interneurones through extracellular electrodes. *Brain Res. 109: 158-164* **vis opt IN tech anat**

KIEN J 1976 Arousal changes in the locust optomotor system. *J. Insect Physiol. 22: 393* **vis IN MN OM**

KIEN J 1977 Comparison of sensory input with motor output in the locust optomotor system. *J. comp. Physiol. A 113: 161-179* **vis IN MN opt neck OM**

KIEN J 1979 Variability of locust motoneuron responses to sensory stimulation: a possible substrate for motor flexibility. *J. comp. Physiol. A 134: 55-68* **MN walk**

KIEN J 1980 Motor control by pleurisegmental interneurons in the locust. *Adv. Physiol. Sci. 23: 515-535* **IN MN thor**

KIEN J 1980 Mechanism of motor control by plurisegmental interneurons in locusts. *J. comp. Physiol. A 140: 303-320* **IN MN thor**

KIEN J 1980 Morphology of locust neck muscle motoneurons and some of their inputs. *J. comp. Physiol. A 140: 321-336* **anat thor MN DN SO neck**

KIEN J 1983 The initiation and maintenance of walking in the locust: an alternative to the command concept. *Proc. R. Soc. Lond. (Biol.) 219: 137-174* **DN walk**

KIEN J 1990 Neuronal activity during spontaneous walking - I. Starting and stopping. *Comp. Biochem. Physiol. A 95: 607-621* **walk IN MN thor**

KIEN J 1990 Neuronal activity during spontaneous walking - II. Correlation with stepping. *Comp. Biochem. Physiol. A 95: 623-638* **walk IN MN thor**

KIEN J ALTMAN JS 1979 Connections of the locust wing tegula with metathoracic flight motoneurons. *J. comp. Physiol. A 133: 299-310* **flight SO teg MN propr**

KIEN J ALTMAN JS 1984 Descending interneurones from the brain and suboesophageal ganglion and their role in control of locust behaviour. *J. Insect Physiol. 30: 59-72* **IN brain suboes DN walk oc**

KIEN J FLETCHER WA ALTMAN JS RAMIREZ JM ROTH U 1990 Organisation of intersegmental interneurons in the suboesophageal ganglion of *Schistocerca gregaria* (Forskål) and *Locusta migratoria migratorioides* (Reiche & Fairmaire) (Acrididae, Orthoptera). *Int. J. Insect Morphol. Embryol. 19: 35-60.* **suboes IN anat**

KIEN J LAND MF 1978 The fast phase of optokinetic nystagmus in the locust. *Physiol. Entomol. 3: 53-57* **OM vis**

KIEN J WILLIAMS M 1983 Morphology of neurons in locust brain and suboesophageal ganglion involved in initiation and maintenance of walking. *Proc. R. Soc. Lond. (Biol.) 219: 175-192* **DN walk CF suboes brain anat**

KILBY PM RADDA GK 1991 2-aminoethylphosphonic acid is the main phosphorus compound in locust haemolymph. *Naturwiss. 78: 514-517* **blood metab**

KILLMANN F SCHÜRMANN FW 1985 Both electrical and chemical transmission between the "lobula giant movement detector" and the "descending contralateral movement detector" neurons of locusts are supported by electron microscopy. *J. Neurocytol. 14: 637-652* **DN fineStruc elect brain**

KILLMANN FJ 1989 Zur Typologie, Verteilung und Anzahl synaptischer Kontakte an den dendritischen Verzweigungen eines identifizierten Interneurons im Gehirn der Wanderheuschrecke *Schistocerca gregaria. Diplomarbeit, Universität Göttingen, Germany.* **anat brain DN vis fineStruc**

KIRCHHOF B BICKER G 1990 Locust neurons in cell culture. *In: Elsner N Roth G (eds.) Brain - Perception - Cognition: Proc. 18th Göttingen Neurobiol. Conf. Stuttgart, New York: Thieme Verlag. Abstract 459* **cell-cult**

KIRSCHENBAUM S O'SHEA M 1990 Neurosecretory cells in the locust corpora cardica are born throughout development and maturation. *Soc. Neurosci. Abstr. 20: 633* **devel retroc CC NSC**

KISS T VARANKA I BENEDEKZY I 1984 Neuromuscular transmission in the visceral muscle of locust oviduct. *Neuroscience 12: 309-322* **NM genit**

KITS KS USHERWOOD PNR 1984 Ion selectivity of glutamate gated channels. *In: Cottrell GA (ed.), Electropharmacology of the in vitro synapse. St. Andrews: University Press. 140* **musc NM glu channel TS**

KITS KS USHERWOOD PNR 1988 Ion-selectivity of single glutamate-gated channels in locust skeletal muscle. *J. exp. Biol. 138: 499-516* **musc glu channel NM TS**

KLEE S THURM U 1986 Mechanorezeptive Fadenhaar-Sensillen mit zwei Sinneszellen von *Locusta migratoria* L. *Verh. dt. zool. Ges. 1986: 220* **SO windh mech fineStruc mech**

KLEMM N 1972 Monoamine containing nervous fibres in the foregut and salivary gland of the desert locust *Schistocerca gregaria* Forskål (Orthoptera, Acrididae). *Comp. Biochem. Physiol. A 43: 207-211* **TS gut SGG saliv**

KLEMM N AXELSSON S 1973 Detection of dopamine, noradrenaline and 5-hydroxytryptamine in the cerebral ganglion of the desert locust *Schistocerca gregaria* Forsk. (Insecta: Orthoptera). *Brain Res. 57: 289-298* **TS brain 5HT norad dopa**

KLEMM N FALCK B 1978 Monoamines in the pars intercerebralis-corpus cardiacum complex of locusts. *Gen. comp. Endocrinol. 34: 180-192* **TS brain retroc**

KLEMM N HUSTERT R CANTERA R NÄSSEL DR 1986 Neurons with reactive antibodies against serotonin in the stomatogastric nervous system and in the alimentary canal of locust and crickets (Orthoptera, Insecta). *Neuroscience 17: 247-261* **TS 5HT SGG gut gryll**

KLEMM N SCHNEIDER L 1975 Selective uptake of indolamine into nervous fibres in the brain of the desert locust *Schistocerca gregaria* Forskål (Insecta). A fluorescence and electron microscopic investigation. *Comp. Biochem. Physiol. C 50: 177-182* **TS brain**

KLEMM N SUNDLER F 1983 Organization of catecholamine and serotin-immunoreactive neurons in the corpora pedunculata of the desert locust, *Schistocerca gregaria* Forsk. *Neurosci. Lett. 36: 13-17* **TS brain 5HT**

KLINGER S RONACHER B 1991 GABA-like immunoreactivity of identified auditory interneurons of the locust. *In: Elsner N Penzlin H (eds.), Synapse-transmission-modulation: Proc. 19th Göttingen Neurobiol. Conf.*

Stuttgart, New York: Thieme Verlag. Abstract 139 **TS GABA anat IN aud**

KLOSE M BENTLEY D 1989 Transient pioneer neurons are essential for the formation of an embryonic peripheral nerve. *Science 245: 982-984* **devel**

KNETSCH H 1939 Die Korrelation in der Ausbildung der Tympanalorgane, der Flügel, der Stridulationsapparate und andere Organsysteme bei den Orthopteren. *Archs. Naturgesch. (N.F.) 8: 1-69* **SO tymp strid wing anat aud**

KNIPPER M BOEKHOFF I BREER H 1989 Isolation and reconstitution of the high-affinity choline carrier. *FEBS Lett. 245: 235-237* **TS ACh metab**

KNIPPER M BREER H 1986 Regulation of the high affinity choline uptake in locust synaptosomes by adenosine triphosphate. *Neurosci. Lett. 72: 347-351* **synaptosome TS ACh**

KNIPPER M BREER H 1987 Protein kinase C in the nervous system of insects: effects of phorbol esters on cholinergic synapses. *Neurochem. Int. 10: 323-328* **TS ACh enz**

KNIPPER M BREER H 1988 Subtypes of muscarinic receptors in insect nervous system. *Comp. Biochem. Physiol. C 90: 275-280* **TS ACh recept**

KNIPPER M BREER H 1989 Muscarinic receptors modulating acetylcholine release from insect synaptosomes. *Comp. Biochem. Physiol. C 93: 287-293* **TS ACh recept synaptosome mod**

KNIPPER M KRIEGER J BREER H 1989 Hemicholinum-3 binding sites in the nervous tissue of insects. *Neurochem. Int. 14: 211-215* **TS ACh**

KNIPPER M MADEDDO L BREER H MELDOLESI J 1986 Black widow spider venom-induced release of neurotransmitters from synaptosomes: a unique venom component (latrotoxin) is active on mammalian, multiple components on insect nerve terminals. *Neurosci. 19: 55-63* **tox synaptosome TS**

KNIPPER M STROTMANN J MÄDLER U KAHLE C BREER H 1989 Monoclonal antibodies against the high affinity choline transport system. *Neurochem. Int. 14: 217-222* **TS ACh**

KNYAZEVA NI 1974 [The campaniform sensilla of *Locusta migratoria* L. (Orthoptera, Acrididae).] (in Russian.) *Entomol. Obozr. 53: 91-95* **SR propr wing leg thor ant**

KNYAZEVA NI 1986 [Morphofunctional features of stretch receptors and chordotonal organs in the wing apparatus of the locust *Locusta migratoria]*. (in Russian.) *Zh. evol. Biokhim. Fiziol. 22: 47-54.* **SR CO SO anat flight propr**

KNYAZEVA NI 1989 [Mechanoreceptors of the head in *Acrida oxycephala*]. (in Russian.) *Zh. evol. Biokhim. Fiziol. 25: 245-48.* **SO windh mech head**

KNYAZEVA NI 1990 Mechanoreceptors and peculiarities of locomotor behaviour of insects. *In: Gribakin FG Wiese K Popov AV (eds.), Sensory systems and communication in arthropods. Basel, Boston, Berlin: Birkhäuser Verlag. 259-264* **windh SO mech leg flight wing**

KOCH UT 1977 A miniature movement detector applied to recording of wingbeat in *Locusta. Fortschr. Zool. 24: 327-332* **flight tech**

KOLLER P FATHI M REICHERT H 1989 Serotonergic neuromodulation of the locust wing stretch receptor. *In: Erber J Menzel R Pflüger H-J Todt D (eds.), Neural mechanisms of behavior: Proc. 2nd Internat. Congr. Neuroethology. Stuttgart & New York: Thieme Verlag. Abstract 233* **SR mod 5HT pep flight SO**

KOLLER P FATHI M REICHERT H 1990 Serotoninergic neuromodulation of the wing stretch receptor in *Schistocerca gregaria* and *Locusta migratoria. In: Elsner N Roth G (eds.) Brain - Perception - Cognition: Proc. 18th Göttingen Neurobiol. Conf. Stuttgart, New York: Thieme Verlag. Abstract 76* **5HT SR flight SO mod**

KONG KL FUNG YM WASSERMAN GS 1980 Filter-mediated color vision with one visual pigment. *Science 207: 783-786* **vis ret**

KONINGS PNM VULLINGS HGB GEFFARD M BUIJS RM DIEDEREN JHB JANSEN WF 1988 Immunocytochemical demonstration of octopamine-immunoreactive cells in the nervous system of *Locusta migratoria* and *Schistocerca gregaria. Cell Tiss. Res. 251: 371-379* **TS oct retroc thor brain DUM VUM opt anat**

KONINGS PNM VULLINGS HGB KOK OJM DIEDEREN JHB JANSEN WF 1989 The innervation of the corpus cardiacum of *Locusta migratoria*: a neuroanatomical study with the use of lucifer jellow. *Cell Tiss. Res. 258: 301-308* **retroc anat CC innerv brain**

KONINGS PNM VULLINGS HGB SIEBINGA R DIEDEREN JHB JANSEN WF 1988 Serotonin-immunoreactive neurones in the brain of *Locusta migratoria* innervating the corpus cardiacum. *Cell Tiss. Res. 254: 147-153* **TS 5HT retroc brain SGG CC anat**

KONINGS PNM VULLINGS HGB VAN GEMERT WMJB DE LEEUW R DIEDEREN JHB JANSEN WF 1989 Octopamine-binding sites in the brain of *Locusta migratoria. J. Insect Physiol. 35: 519-524* **TS oct recept brain**

KOTRLA KJ GOODMAN CG 1984 Transient expression of a surface antigen on a small subset of neurones during embryonic development. *Nature 311: 151-153* **devel MN thor**

KRAHE R STUMPNER A RONACHER B 1991 Reactions of tympanal receptor fibres of the locust to artificial sound stimuli. *In: Elsner N Penzlin H (eds.), Synapse-transmission-modulation: Proc. 19th Göttingen Neurobiol. Conf. Stuttgart, New York: Thieme Verlag. Abstract 140* **aud tymp SO**

KRIEGER J RAMING K KNIPPER M GRAU M MERTENS S BREER H 1990 Cloning, sequencing and expression of locust tropomyosin. *Insect Biochem. 20: 173-184* **musc**

KROGH IM NORMANN TC 1977 The corpus cardiacum neurosecretory cells of *Schistocerca gregaria.* Electron microscopy of resting and secreting cells. *Acta. Zool. (Stockh.) 58: 69-79* **anat CC NSC retroc fineStruc**

KRÜPPEL T GEWECKE M 1985 Visually induced flight manoeuvres in the tethered locust (*Schistocerca gregaria*). *In: Gewecke M Wendler G (eds.), Insect Locomotion. Hamburg & Berlin: Paul Parey Verlag. 167-174* **flight OM**

KÜHNE R 1982 Neurophysiology of the vibratory sense in locusts and bushcrickets: response characteristics of single receptor units. *J. Insect Physiol. 28: 155-163* **vibr SO subg tettig**

KÜHNE R 1982 Neurophysiology of the vibration sense in locusts and bushcrickets: the responses of ventral cord neurons. *J. Insect Physiol. 28: 615-623* **aud vibr IN thor tettig**

KUSTER JE FRENCH AS 1983 Sensory transduction in a locust multipolar joint receptor: the dynamic behaviour under a variety of stimulus conditions. *J. comp. Physiol. A 150: 207-215* **propr SO leg**

KUTSCH W 1970 Nervöse Kontrolle der Aktivität eines thorakalen Muskels der Heuschrecke *Schistocerca gregaria* Forskål. *Zool. Anz. (Suppl.) 33: 489-493* **flight MN**

KUTSCH W 1971 The development of the flight pattern in the desert locust, *Schistocerca gregaria. Z. vergl. Physiol. 74: 156-168* **flight devel**

KUTSCH W 1973 The influence of age and culture-temperature on the wing-beat frequency of the migratory locust, *Locusta migratoria. J. Insect Physiol. 19: 763-772* **flight devel temp**

KUTSCH W 1974 The influence of the wing sense organs on the flight motor pattern in maturing locusts. *J. comp. Physiol. 88: 413-424* **flight SO SR devel propr**

KUTSCH W 1976 Postlarval development of two rhythmical behavioural patterns: flight and song in the grasshopper, *Omocestus viridulus. Physiol. Entomol. 1: 255-263* **devel rhythm strid flight**

KUTSCH W 1983 Aspects of the ontogeny of locust flight. *In: Nachtigall W (ed.), BIONA-report 2 "Insekten Flug"; Akad. Wiss. Mainz. Stuttgart & New York: Gustav Fischer Verlag. 71-78* **flight devel**

KUTSCH W 1985 Pre-imaginal flight motor pattern in *Locusta. J. Insect Physiol. 31: 581-586* **devel rhythm flight**

KUTSCH W 1989 Formation of the receptor system in the hindlimb of the locust embryo. *W. Roux's Arch. Dev. Biol. 198: 39-47* **devel SO leg**

KUTSCH W BENTLEY D 1987 Programmed death of peripheral pioneer neurons in the grasshopper embryo. *Dev. Biol. 123: 517-525.* **devel thor leg**

KUTSCH W GEWECKE M 1979 Development of flight behaviour in maturing adults of *Locusta migratoria* II. Aerodynamic parameters. *J. Insect Physiol. 25: 299-304* **flight devel aero**

KUTSCH W HANLOSER H REINECKE M 1980 Light- and electron microscopic analysis of a complex sensory organ: the tegula of *Locusta migratoria. Cell Tiss. Res. 210: 461-478* **flight SO teg propr fineStruc anat**

KUTSCH W MURDOCK LL 1973 Locust flight oscillator: changes in motor output by certain pharmacological agents. *J. Insect Physiol. 19: 1519-1525* **flight TS ACh**

KUTSCH W NEUBAUER K KRÄMER H 1991 Flight motor pattern in abdominal segments of locusts. *J. exp. Biol. 156: 629-634* **flight abd MN**

KUTSCH W SCHNEIDER H 1987 Histological characterization of neurones innervating functionally different muscles of *Locusta. J. comp. Neurol. 261: 515* **flight anat MN musc**

KUTSCH W STEVENSON P 1981 Time-correlated flights of juvenile and mature locusts: a comparison between free and tethered animals. *J. Insect Physiol. 27: 455-459* **flight**

KUTSCH W STEVENSON P 1984 Manipulation of the endocrine system of *Locusta* and the development of the flight motor pattern. *J. comp. Physiol. A 155: 129-138* **flight devel retroc**

KUTSCH W STEVENSON PA 1988 Demonstration of the adult-specific flight motor pattern in isolated ganglia and all instars of *Locusta*, by use of octopamine. *In: Salánki J S-Rózsa K (eds.), Neurobiology of invertebrates (Symp. Biol. Hung. 36). Budapest. 545-568* **flight TS oct mod**

KUTSCH W USHERWOOD PNR 1970 Studies of the innervation and electrical activity of flight muscles in the locust, *Schistocerca gregaria. J. exp. Biol. 52: 299-312* **flight anat MN musc innerv**

KUWADA JY GOODMAN CS 1985 Neuronal determination during embryonic development of the grasshopper nervous system. *Dev. Biol. 110: 114-126* **devel**

LA GRECA M 1947 Morfologia funzionale dell'articolazione alare degli Ortotteri. *Arch. zool. ital. 32: 271-327* **flight thor anat**

LA GRECA M 1949 I nervi motori e sensori degli organi di volo degli Acrididi. *R. C. Accad. Napoli 15: 121-30* **flight thor anat**

LAFON-CAZAL M 1981 Monoamines in the corpora cardiaca of locusts. *Adv. Physiol. Sci. 22: 255-265* **TS retroc CC**

LAFON-CAZAL M ARLUISON M 1976 Localization of monoamines in the corpora cardiaca and the hypocerebral ganglion of locusts. *Cell Tiss. Res. 72: 517-527* **TS retroc SGG CC**

LAFON-CAZAL M BAEHR JC 1988 Octopaminergic control of corpora allata activity in an insect. *Experientia 44: 895-896* **TS oct retroc CAll**

LAFON-CAZAL M COULON JF DAVIS JC 1982 Octopamine metabolism in the cephalic ganglion of *Locusta migratoria* L. *Comp. Biochem. Physiol. C 73: 293-296* **TS oct brain metab**

LAGUEUX M LWOFF L MEISTER M GOLTZENE F HOFFMANN JA 1990 cDNAs from neurosecretory cells of brains of *Locusta migratoria* (Insecta, Orthoptera) encoding a novel member of the superfamily of insulins. *Eur. J. Biochem. 187: 249-254* **NSC brain pep cDNA insul**

LAIRD AK 1943 A study of the types of male gonads in the Acrididae (Orth.). *J. Morphol. 72: 477-490* **anat genit**

LAKES R 1988 Postembryonic determination and plasticity in the auditory system of *Locusta migratoria. Monographs in Dev. Biol. 21: 214-221* **aud regen**

LAKES R 1989 Specificity of regenerated tympanic fibres in locusts. *In: Elsner N Singer W (eds.), Dynamics and plasticity in neuronal systems: Proc. 17th. Göttingen Neurobiology Conference. Stuttgart, New York: Thieme Verlag. Abstract 4* **tymp SO aud regen anat IN**

LAKES R 1990 Plasticity of the nervous system of orthopterans. *In: Gribakin FG Wiese K Popov A V (eds.), Sensory systems and communication in arthropods. Basel, Boston, Berlin: Birkhäuser Verlag. 280-284* **regen**

LAKES R KALMRING K 1991 Regeneration of the projection and synaptic connections of tympanic receptor fibers of *Locusta migratoria* (Orthoptera) after axotomy. *J. Neurobiol. 22: 169-181* **regen aud tymp SO**

LAKES R KALMRING K ENGELHARD KH 1990 Changes in the auditory system of locusts (*Locusta migratoria* and *Schistocerca gregaria*) after deafferentation. *J. comp. Physiol. A 166: 553-563* **aud IN SO tymp anat regen**

LAKES-HARLAN R PFAHLERT C 1991 Recovery of auditory behavior after axotomy of the tympanal nerve of a grasshopper. *In: Elsner N Penzlin H (eds.), Synapse-transmission-modulation: Proc. 19th Göttingen Neurobiol. Conf. Stuttgart, New York: Thieme Verlag. Abstract 525* **regen behav SO tymp aud**

LAKES-HARLAN R WEILAND G 1991 Peripheral origin of peripheral ganglia. *In: Elsner N Penzlin H (eds.), Synapse-transmission-modulation: Proc. 19th Göttingen Neurobiol. Conf. Stuttgart, New York: Thieme Verlag. Abstract 513* **devel**

LANE NJ 1968 The thoracic ganglia of the grasshopper *Melanoplus differentialis*: fine structure of the perineurium and neuroglia with special reference to the intracellular distribution of phosphatases. *Z. Zellforsch. mikr. Anat. 86: 293-312* **fineStruc glia enz thor**

LANE NJ TREHERNE JE 1973 The ultrastructural organization of peripheral nerves in two insect species (*Periplaneta americana* and *Schistocerca gregaria*). *Tissue & Cell 5: 703-714* **fineStruc glia dict**

LANG F 1988 The influence of hindleg movement on the auditory system of *Locusta migratoria. In: Elsner N Barth FG (eds.) Sense organs: interfaces between environment and behaviour: Proc.16th. Göttingen Neurobiol. Conf. Stuttgart, New York: Thieme Verlag. Abstract 25* **aud SO mech IN thor**

LANG F ELSNER N 1989 The interference of sound and movement stimuli in auditory interneurons of *Locusta migratoria. J. comp. Physiol. A 164: 697-706* **aud mech thor IN strid**

LANGE AB ORCHARD I 1984 Some pharmacological properties of neuromuscular transmission in the oviduct of the locust, *Locusta migratoria. Archs. Insect Biochem. Physiol. 1: 231-241* **NM genit**

LANGE AB ORCHARD I 1984 Neural inhibition of egglaying in the locust, *Locusta migratoria. J. Insect Physiol. 30: 271-278* **genit inhIB**

LANGE AB ORCHARD I 1984 Dorsal unpaired median neurones and ventral bilaterally paired neurones project to a visceral muscle in an insect. *J. Neurobiol. 15: 441-453* **DUM genit anat MN abd**

LANGE AB ORCHARD I 1986 Identified octopaminergic neurons modulate contractions of locust visceral muscle via adenosine 3′,5′-monophosphate (cyclic AMP). *Brain Res. 363: 340* **NM oct cAMP genit DUM**

LANGE AB ORCHARD I 1986 Ventral neurons in an abdominal ganglion of the locust with properties similar to dorsal unpaired median neurons. *Can. J. Zool. 64: 264-266* **MN abd anat**

LANGE AB ORCHARD I 1990 The action of phenyliminoimidazolines and 2-aminooxazoline on octopamine receptors on locust fat body. *Pest. Biochem. Physiol. 37: 24-29* **TS oct recept tox**

LANGE AB ORCHARD I ADAMS ME 1986 Peptidergic innervation of insect reproductive tissue: the association of proctolin with oviduct visceral musculature. *J. comp. Neurol. 254: 279-286* **pep NM TS musc genit proct**

LANGE AB ORCHARD I LAM W 1987 Mode of action of proctolin on locust visceral muscle. *Archs. Insect Biochem. Physiol. 5: 285-295* **pep NM TS musc genit proct**

LANGE AB ORCHARD I LOUGHTON BG 1984 Spontaneous and neurally evoked contractions of visceral muscles in the oviduct of *Locusta migratoria. Archs. Insect Biochem. Physiol. 1: 179-190* **genit musc**

LANGE AB ORCHARD I TE BRUGGE VA 1991 Evidence for the involvement of a SchistoFLRF-amide-like peptide in the neural control of locust oviduct. *J. comp. Physiol. A 168: 383-391* **genit pep FMRF musc mod**

LAUGA J ALBRECHT FO CASANOVA D 1981 Influence of rearing humidity and photoperiod on the locomotor and flight activities of solitarious migratory locusts *Locusta migratoria migratorioides* (R. & F.) (Orthoptera: Acrididae). *Acrida 10: 219-232* **flight walk behav**

LAUGHLIN SB LILLYWHITE PG 1982 Intrinsic noise in locust photoreceptors. *J. Physiol. (Lond.) 332: 25-45* **ret**

LAURENT G 1986 Thoracic intersegmental interneurones in the locust with mechanoreceptive inputs from a leg. *J. comp. Physiol. A 159: 171-186* **IN leg thor mech**

LAURENT G 1987 The morphology of a population of thoracic intersegmental interneurones in the locust. *J. comp. Neurol. 256: 412-429* **IN leg anat thor**

LAURENT G 1987 Parallel effects of joint receptors on motor neurones and inter-segmental interneurones in the locust. *J. comp. Physiol. A 160: 341-353* **MN IN SO leg mech**

LAURENT G 1987 The role of spiking local interneurons in shaping the receptive fields of intersegmental interneurons in the locust. *J. Neurosci. 7: 2977* **IN thor leg**

LAURENT G 1988 Local circuits underlying excitation and inhibition of intersegmental interneurons in the locust. *J. comp. Physiol. A 162: 145-157* **local IN leg thor**

LAURENT G 1990 Intersegmental interneurones and interactions between centres of local integration in the locust central nervous system. *In: Armstrong D Bush BMH (eds.), Mechanisms of arthropod and vertebrate locomotion. Manchester: Manchester University Press. 11-24* **IN MN leg thor nonspik**

LAURENT G 1990 Voltage-dependent non-linearities in the membrane of locust nonspiking local interneurones, and their significance for synaptic integration. *J. Neurosci. 10: 2268-2280* **IN thor nonspik**

LAURENT G 1991 Evidence for voltage-activated outward currents in the neuropilar membrane of locust nonspiking local interneurons. *J. Neurosci. 11: 1713-1726* **IN nonspik thor**

LAURENT G 1991 The variability of transmitter release probability at central synapses between local interneurons in the locust. *Soc. Neurosci. Abstr. 17: 578* **thor IN local TS nonspik**

LAURENT G BURROWS M 1988 A population of ascending intersegmental interneurones in the locust with mechanosensory inputs from a hind leg. *J. comp. Neurol. 275: 1-12* **IN thor leg anat FCO propr**

LAURENT G BURROWS M 1988 Direct excitation of nonspiking local interneurones by exteroceptors underlies tactile reflexes in the locust. *J. comp. Physiol. A 162: 563-572* **SO nonspik leg thor IN anat**

LAURENT G BURROWS M 1989 Distribution of intersegmental inputs to nonspiking local interneurones and motor neurones in the locust. *J. Neurosci. 9: 3019-3029* **IN MN leg thor nonspik**

LAURENT G BURROWS M 1989 Intersegmental interneurons can control the gain of reflexes in adjacent segments of the locust by their action on nonspiking local interneurons. *J. Neurosci. 9: 3030-3039* **IN MN leg thor nonspik**

LAURENT G HUSTERT R 1988 Motor neuronal receptive fields delimit the patterns of motor activity during locomotion of the locust. *J. Neurosci. 8: 4349-4366* **MN thor leg SO anat**

LAVENSEAU L FOURNIER B GIRARDIE J GOURDOUX L MOREAU R PEYPELUT L PICQUOT M PROUX J PUIROUX J GIRARDIE A 1991 Endocrine regulation of hydromineral and energetic metabolisms under extreme environmental conditions in insects. *In: Loughton BG, Saluddin ASM (eds.) Neurobiology and endocrinology of invertebrates. Toronto: York University Press, Inc. 115-144* **hormX metab**

LAZAROVICI P PENER MP 1978 The relations of the pars intercerebralis, corpora allata, and juvenile hormone to oocyte development and oviposition in the african migratory locust. *Gen. comp. Endocrinol. 35: 375-386* **brain retroc**

LE BERRE JR LOUVEAUX A 1969 Equipement sensorial des mandibules de la larve du premier stade de *Locusta migratoria* L. *C. R. hebd. Seanc. Acad. Sci. Paris Ser D 268: 2907-2910* **SO mpt anat**

LE BERRE JR SINOIR Y BOULAY C 1967 Etude de l'equipement sensoriel de l'article distal des palpes chez la larve de *Locusta migratoria migratorioides* (R. & F.). *C. R. hebd. Seanc. Acad. Sci. Paris Ser D 265: 1717-1720* **SO mpt anat**

LEA TJ USHERWOOD PNR 1970 Increased chloride permeability of insect muscle fibres on exposure to ibotenic acid. *J. Physiol. (Lond.) 212: 32-33P* **TS glu NM channel**

LEA TJ USHERWOOD PNR 1973 The site of action of ibotenic acid and the identification of two populations of glutamate receptors on insect muscle-fibres. *Comp. gen. Pharmacol. 4: 333-350* **TS glu recept musc**

LEA TJ USHERWOOD PNR 1973 Effect of ibotenic acid on chloride permeability of insect muscle-fibres. *Comp. gen. Pharmacol. 4: 351-363* **TS glu musc channel**

LECHLEITNER RA AUDSLEY N PHILIPS JE 1989 Antidiuretic action of cAMP, corpus cardiacum and ventral ganglia on fluid absorption across locust ileum in vitro. *Can. J. Zool. 67: 2655-2661* **gut pep hormX retroc CC abd cAMP**

LEE RM 1961 The variation of blood volume with age in the desert locust (*Schistocerca gregaria* Forsk.). *J. Insect Physiol. 6: 36-51* **blood tech**

LEECH CA 1986 Resting potential and potassium-selective electrode measurements in locust skeletal msucles. *J. exp. Biol. 122: 439-442* **musc tech**

LEES G BEADLE DJ NEUMANN R BENSON JA 1987 Responses to GABA by isolated neuronal somata: pharmacology and modulation by a benzodiazepine and a barbiturate. *Brain Res. 401: 267-278* **cellCult TS GABA**

LEES G BURT PE 1988 Neurotoxic actions of a lipid amide on the cockroach nerve cord and on locust somata maintained in short-term culture: a novel preparation for the study of Na+ channel pharmacology. *Pest. Sci. 24: 189-191* **cellCult channel dict tox lip**

LEES G PEARSON HA WRAY D 1989 Calcium channels in isolated locust thoracic ganglia. *J. Physiol. (Lond.) 410: 13P* **channel cellcult**

LEFCORT F BENTLEY D 1985 Neuronal guidance in dissected limb buds, in cell culture and at ectopic locations in host limb buds. *Soc. Neurosci. Abstr. 11: 334* **devel cellCult leg**

LEFCORT F BENTLEY D 1987 Pathfinding by pioneer neurons in isolated, opened and mesoderm-free limb buds of embryonic grasshoppers. *Dev. Biol. 119: 466-480* **devel leg**

LEFCORT F BENTLEY D 1987 Intrinsic cell polarity and orientation of afferent pioneer axon outgrowth in the grasshopper embryo. *Soc. Neurosci. Abstr. 13: 1221* **devel leg**

LEFCORT F BENTLEY D 1989 Organization of cytoskeletal elements and organelles preceding growth cone emergence from an identifed neuron in situ. *J. Cell Biol. 108: 1737-1749* **devel leg**

LEITCH B LAURENT G SHEPHERD D 1991 Synaptogenesis in identified spiking local interneurons of the locust *Schistocerca gregaria. Soc. Neurosci. Abstr. 17: 217* **fineStruc IN thor local dev**

LEMMNITZ G 1987 Yaw-Bewegung durch optisch induzierte Steuermanöver beim Heuschreckenflug. *In: Elsner N Creutzfeld O (eds.), New frontiers in brain research: Proc. 15th Göttingen Neurobiol. Conf. Stuttgart, New York: Thieme Verlag. Abstract 49* **flight OM**

LEMMNITZ G 1988 Optically-induced activity in thoracic neurons of tethered flying locusts. *In: Elsner N Barth FG (eds.) Sense organs: interfaces between environment and behaviour: Proc.16th. Göttingen Neurobiol. Conf. Stuttgart, New York: Thieme Verlag. Abstract 127* **flight IN thor vis**

LETOURNEAU PC 1982 Nerve fiber growth and its regulation by extrinsic factors. *In: Spitzer NC (ed.), Neuronal development. New York: Plenum Press. 213-254* **devel**

LEWIS GW MILLER PL MILLS PS 1973 Neuromuscular mechanisms of abdominal pumping in the locust. *J. exp. Biol. 59: 149-168* **resp anat MN innerv**

LILLYWHITE PG 1977 Single photon signals and transduction in an insect eye. *J. comp. Physiol. A 122: 189-200* **ret**

LILLYWHITE PG 1978 Coupling between locust photoreceptors revealed by a study of quantum bumps. *J. comp. Physiol. A 125: 13-27* **ret**

LILLYWHITE PG 1981 A new model for phototransduction. *Nature 290: 358-359* **ret**

LILLYWHITE PG LAUGHLIN SB 1979 Transducer noise in a photoreceptor. *Nature 277: 569-571* **ret**

LINDBERG D ELSNER N 1977 Sensory influence upon grasshopper stridulation. *Naturwiss. 64: 342* **strid**

LINK E 1909 Über die Stirnaugen der hemimetabolen Insekten. *Zool. Jb. (Anat.) 27: 281-376* **oc anat**

LINS F ELSNER N 1990 Descending interneurons of the suboesophageal ganglion that inhibit stridulation in acridid grasshoppers. *In: Elsner N Roth G (eds.) Brain - Perception - Cognition: Proc. 18th Göttingen Neurobiol. Conf. Stuttgart, New York: Thieme Verlag. Abstract 115* **suboes IN strid inhib**

LINS F LAKES-HARLAN R ELSNER N 1991 GABAergic immunoreactivity and physiology of descending interneurons in the stridulatory system of grasshoppers. *In: Elsner N Penzlin H (eds.), Synapse-transmission-modulation: Proc. 19th Göttingen Neurobiol. Conf. Stuttgart, New York: Thieme Verlag. Abstract 141* **TS GABA anat IN strid**

LITTLEWOOD PMH SIMMONS PJ 1988 Hexamminecobaltic chloride provides a simple method for marking neurones for electron microscopy. *Brain Res. 445: 165-170* **oc IN brain anat tech**

LITTLEWOOD PMH SIMMONS PJ 1989 Comparative physiology and ultrastructure of tonic and phasic synapses between identified neurones in the insect ocellar tract. *In: Erber J Menzel R Pflüger H-J Todt D (eds.), Neural mechanisms of behavior: Proc. 2nd Internat. Congr. Neuroethology. Stuttgart & New York: Thieme Verlag. Abstract 176* **oc IN brain fineStruc**

LIU WEI-TEH 1952 [Über die Morphologie und Entwicklung des Abdominaltympanalorgans der Heuschrecke (*Locusta migratoria manilensis* Meyen) (Acrididae, Orthoptera)]. (in Chinese, with German summary). *Acta entomol. Sin. 2: 155-65* **SO tymp aud anat devel**

LOER CM STEEVES JD GOODMAN CS 1983 Neuronal cell death in grasshopper embryos: variable patterns in different species, clutches, and clones. *J. Embryol. exp. Morphol. 78: 169-182* **devel**

LOHER W 1958 An olfactory response of immature adults of the desert locust. *Nature 181: 1280.* **chem ant mech behav**

LOMBARDO CA 1973 On the presence of two coxal sense organs in Pterygota insects. *Monit. zool. Ital. (NS) 7: 243-246* **propr leg**

LOUGHTON BG 1987 Studies on locust hypolipaemic hormone. *J. Insect Physiol. 33: 569-580* **pep**

LOUGHTON BG ORCHARD I 1981 The nature of the hyperglycaemic factor from the glandular lobe of the corpus cardiacum of *Locusta migratoria. J. Insect Physiol. 27: 383-385* **pep retroc AKH**

LOUVEAUX A 1972 Equipement sensoriel et système nerveux péripherique des pièces buccales de *Locusta migratoria* L. *Insectes Soc. 19: 359-368* **mpt SO anat**

LOUVEAUX A 1973 Etude d'une plage d'organes sensoriels sur les galea de *Locusta migratoria* L. *C. R. hebd. Seanc. Acad. Sci. Paris Ser D 277: 1355-1356* **feed mpt SO**

LOUVEAUX A 1973 Etude de l'innervation sensorielle de l'hypopharynx de larve de *Locusta migratoria migratorioides* R. & F. (Orthoptère, Acrididae). *Insectes Sociaux 22: 3-12* **mpt SO anat**

LOUVEAUX A 1978 Feeding mechanisms of the migratory locust, *Locusta migratoria migratorioides* (Orthoptera, Acrididae). *Entomol. exp. appl. 24: 514-219* **feed mpt**

LOWENSTEIN O FINLAYSON LH 1960 The response of the abdominal stretch receptor of an insect to phasic stimulation. *Comp. Biochem. Physiol. 1: 56-61* **SO propr abd**

LUH CHIN-JEN YU PEI-YU 1957 [The skeleto-muscular system of the asiatic migratory locust, *Locusta migratoria manilensis* (Meyen). I. The head region| (in Chinese with English summary). *Acta entomol. Sin. 7: 1-19* **anat head musc neck**

LUNT GG 1973 Hydrophobic proteins from locust (*Schistocerca gregaria*) muscle with glutamate receptor properties. *Comp. gen. Pharmacol. 4: 75-79* **glu recept TS**

LUNT GG ROBINSON TR MILLER T KNOWLES WP OLSEN RW 1985 The identification of GABA receptor binding sites in insect ganglia. *Neurochem. Int. 7: 751-754* **TS GABA recept**

LUTZ EM LLOYD SJ TYRER NM 1988 Purification of choline acetyltransferase from the locust *Schistocerca gregaria* and production of serum antibodies to this enzyme. *J. Neurochem. 50: 82-88* **TS Ach enz**

LUTZ EM TYRER NM 1987 Immunohistochemical localization of choline acetyltransferase in the central nervous system of the locust. *Brain Res. 407: 173-179* **TS ACh enz anat**

LUTZ EM TYRER NM 1988 Immunohistochemical localization of serotonin and choline acetyltransferase in sensory neurons of the locust. *J. comp. Neurol. 267: 335-342* **TS 5HT ACh SO anat**

LUTZ EM TYRER NM ALTMAN JS TURNER JD 1985 Some insect sensory neurones contain 5-hydroxytryptamine. *Brain Res. 325: 353-356* **TS 5HT ACh SO**

MACDONALD AG MASON D RAMSEY RL SHELTON C USHERWOOD PNR 1989 Ketamine is a closed and open channel blocker of a locust (*Schistocerca gregaria*) muscle glutamate receptor. (Abstract). *J. Physiol. (Lond.) 417: 86P* **TS glu musc recept tox**

MACDONALD AG RAMSEY RL SHELTON C USHERWOOD PNR 1989 An apparatus for single channel patch recording at high pressure. *J. Physiol. (Lond.) 409: 2P* **channel tech**

MACFARLANE JE 1953 The morphology of the chordotonal organs of the antenna, mouthparts and pegs of the lesser migratory grasshopper, *Melanoplus mexicanus mexicanus* (Saussure). *Can. Entomol. 85: 81-103* **anat CO ant leg mpt SO**

MACFARLANE RG MIDGLEY JM WATSON DG EVANS PD 1990 The analysis of biogenic amines in the thoracic nervous system of the locust, *Schistocerca gregaria*, by gas-chromatography negative ion chemical ionization mass spectrometry (GC-NICIMS). *Insect Biochem. 20: 305-311* **tech thor TS oct dopa norad 5HT tyr**

MACFARLANE RG MIDGLEY JM WATSON DG EVANS PD 1990 Identification and quantitation of phenylalanine, tyrosine and dihydroxyphenylalanine in the thoracic nervous system of the locust *Schistocerca gregaria* by gas-chromatography-negative ion chemical ionisation mass spectrometry. *J. Chromatogr. (Biomed Applicns.) 532: 1-11* **TS tech thor**

MACFARLANE RG MIDGLEY JM WATSON DG EVANS PD 1990 Identification and quantitation of N-acetyl metabolites of biogenic amines in the thoracic nervous system of the locust *Schistocerca gregaria* by gas-chromatography-negative ion chemical ionisation mass spectrometry. *J. Chromatogr. (Biomed Applicns.) 532: 13-25* **thor TS tech**

MACFARLANE RG MIDGLEY JM WATSON DG EVANS PD 1991 Biogenic amines: their occurrence, biosynthesis and metabolism in the locust, *Schistocerca gregaria*, by gas-chromatography-negative ion chemical ionisation mass spectrometry. *J. Chromatogr. (Biomed Applicns.) 562: 585-598* **TS tech metab**

MACMILLAN DL KIEN J 1983 Intra- and intersegmental pathways active during walking in the locust. *Proc. R. Soc. Lond. (Biol.) 318: 287-308* **FCO SO propr walk IN**

MAKI T 1938 Studies on the thoracic musculature of insects. *Mem. Fac. Sci. Agric. Taihoku imp. Univ. 24: 1-344* **musc anat thor**

MAKINGS P 1964 "Slifer's patches" and the thermal sense in Acrididae (Orthoptera). *J. exp. Biol. 41: 473-497* **SO temp head**

MALAMUD JG 1989 The tension in a locust flight muscle at varied muscle lengths. *J. exp. Biol. 144: 479-496* **flight musc**

MALAMUD JG JOSEPHSON RK 1991 Force-velocity relationships of a locust flight muscle at different times during a twitch contraction. *J. exp. Biol. 159: 65-88* **flight musc**

MALAMUD JG MIZISIN AP JOSEPHSON RK 1988 The effects of octopamine on contraction kinetics and power output of a locust flight muscle. *J. comp. Physiol. A 162: 827-835* **flight musc TS oct mod**

MAORI J 1961 Innervation of the insect leg muscle. *Acta biol. Hung. 12: 219-230* **leg musc innerv anat**

MARKL H 1969 Verständigung durch Vibrationssignale bei Arthropoden. *Naturwiss. 56: 499-505* **aud vibr strid**

MARQUART V 1985 Auditorische Interneurone im thorakalen Nervensystem von Heuschrecken: Morphologie, Physiologie und synaptische Verbindungen. *Dissertation, Universität Bochum, Germany.* **aud IN thor anat**

MARQUART V 1985 Local interneurons mediating excitation and inhibition on to ascending neurons in the auditory pathway of grasshoppers. *Naturwiss. 72: 42-44* **aud IN anat local**

MARSHALL J BUCKINGHAM SD SHINGAI R LUNT GG GOOSEY MW DARLINSON MG SATTELLE DB BARNARD EA 1991 Sequence and functional expression of a single α-subunit of an insect nicotinic acetylcholine receptor. *EMBO J. 9: 4391-4398* **brain ACh recept**

MARSHALL J DARLISON MG LUNT GG BARNARD EA 1988 Cloning of putative nicotinic acid acetylcholine receptor genes from the locust. *Biochem. Soc. Trans. 16: 463-465* **ACh TS recept**

MASON CA 1973 New features of the brain-retrocerebral neuroendocrine complex of the locust, *Schistocerca vaga. Z. Zellforsch. mikr. Anat. 141: 19-32* **retroc NSC anat brain CAll CC**

MASON CA O'SHEA M 1979 Recent developments in intracellular staining. *Trends in Neurosci. 2: 69-76* **tech**

MASON JB 1969 The tympanal organ of Acridomorpha. *Eos, Madrid 44: 267-355* **aud SO tymp strid anat rom pamph pyrg rev**

MATHERS D USHERWOOD PNR 1976 Concanavalin A blocks desensitization of glutamate receptors of insect muscle fibres. *Nature 259: 409-411* **musc glu NM desens recept**

MATHERS D USHERWOOD PNR 1978 Effects of concanavalin A on junctional and extrajunctional L-glutamate receptors on locust skeletal muscle fibres. *Comp. Biochem. Physiol. C 59: 151-155* **musc glu NM recept**

MATHERS DA USHERWOOD PNR 1978 The sensitivity of locust skeletal muscle fibres to L-glutamate following denervation and injury. *Comp. Biochem. Physiol. C 60: 7-10* **TS glu musc degen**

MATHESON T 1990 Responses and locations of neurones in the locust metathoracic femoral chordotonal organ. *J. comp. Physiol. A 166: 915-927* **FCO propr anat thor leg SO mech innerv**

MATHESON T 1992 Morphology of the central projections of physiologically characterised neurones from the locust metathoracic femoral chordotonal organ. *J. comp. Physiol. 170: 101-120* **FCO anat thor SO propr**

MATHESON T FIELD LH 1990 Innervation of the metathoracic femoral chordotonal organ of *Locusta migratoria. Cell Tiss. Res. 259: 551-560* **SO propr FCO anat thor leg**

MATIC T LAUGHLIN SB 1981 Changes in the intensity-response function of an insect's photoreceptors due to light adaptation. *J. comp. Physiol. A 145: 169-177* **ret**

MAURUS P AMMERMÜLLER J 1989 Spectral sensitivities of photoreceptors and second order neurons in the ocellar system of *Locusta migratoria. In: Elsner N Singer W (eds.), Dynamics and plasticity in neuronal systems: Proc. 17th. Göttingen Neurobiology Conference. Stuttgart, New York: Thieme Verlag. Abstract 99* **oc ret IN brain**

MAY TE BROWN BE CLEMENTS AN 1979 Experimental studies upon a bundle of tonic fibres in the locust extensor tibialis muscle. *J. Insect Physiol. 25: 169-181* **musc**

MAYER RJ CANDY DJ 1967 Changes in haemolymph lipoproteins during locust flight. *Nature 215: 987* **flight blood metab**

MAYER RJ CANDY DJ 1969 Changes in energy reserves during flight of the desert locust, *Schistocerca gregaria. Comp. Biochem. Physiol. 31: 409-418* **metab flight**

MAYER RJ CANDY DJ 1969 Control of haemolymph lipid concentration during locust flight: an adipokinetic hormone from the corpus cardiacum. *J. Insect Physiol. 15: 611-620* **pep AKH flight retroc blood CC metab**

MAZOUÉ H BUSNEL M-C CHAUCHARD P 1957 Les modifications réflexes de l'excitabilité nerveuse par les sons et les ultrasons chez divers Orthoptéres. *C. R. Acad. Sci. Paris Ser. D 244: 666-668* **aud behav HF flight**

McCAFFERY AR 1972 Relationship of the corpus allatum and the neurosecretory system in locusts. *J. Endocrinol. 57: li-lii* **retroc NSC brain CAll**

McCAFFERY AR HIGHNAM KC 1975 Effects of corpora allata on the activity of the cerebral neurosecretory system of *Locusta migratoria migratorioides. Gen. comp. Endocrinol. 25: 358-372* **retroc NSC brain CAll**

McCAFFERY AR HIGHNAM KC 1975 Effects of corpus allatum hormone and its mimics on the activity of the cerebral neurosecretory system of *Locusta migratoria migratorioides* R. & F. *Gen. comp. Endocrinol. 25: 373-386* **retroc NSC brain CAll JH**

MCINDOO NE 1920 The olfactory sense of Orthoptera. *J. comp. Neurol. 31: 405-427* **SO chem mech anat**

MCKINLAY RG USHERWOOD PNR 1973 The role of the synaptic vesicles in transmission at the insect nerve-muscle junction. *Life Sciences 13: 1051-1056* **NM**

MCKINLAY RG USHERWOOD PNR 1978 The effect of magnesium ions on the fine structure of the insect neuromuscular junction. *J. Ultrastruct. Res. 62: 83-93* **musc fineStruc NM**

MEANS OW 1942 Cholinesterase activity of tissues of adult *Melanoplus differentialis* (Orthoptera, Acrididae). *J. cell. comp. Physiol. 20: 319-324* **ACh TS enz**

MÉDIONI J 1961 Le problème de l'existence d'un sens dermatoptique chez les insectes. L'etat actuel del la question. *Psychol. franç. 6: 11-20* **vis rev**

MEIER T CHABAUD F REICHERT H 1991 Homologous patterns in the embryonic development of the peripheral nervous system in the grasshopper *Schistocerca gregaria* and the fly *Drosophila melanogaster. Development 112: 241-253* **devel**

MEIER T FRIEDLI M REICHERT H 1990 Molecular labels in the neuronal development of the grasshopper. *In: Elsner N Roth G (eds.) Brain - Perception - Cognition: Proc. 18th Göttingen Neurobiol. Conf. Stuttgart, New York: Thieme Verlag. Abstract 369* **devel DN**

MEIER T REICHERT H 1989 Development and evolution of the locust auditory system. *In: Elsner N Singer W (eds.), Dynamics and plasticity in neuronal systems: Proc. 17th. Göttingen Neurobiology Conference. Stuttgart, New York: Thieme Verlag. Abstract 134* **SO tymp CO subg homol devel abd**

MEIER T REICHERT H 1989 Segmentally specialized sensory systems share a common developmental origin in the locust embryo. *In: Erber J Menzel R Pflüger H-J Todt D (eds.), Neural mechanisms of behavior: Proc. 2nd Internat. Congr. Neuroethology. Stuttgart & New York: Thieme Verlag. Abstract 16* **devel tymp CO SR thor abd homol**

MEIER T REICHERT H 1990 Embryonic development and evolutionary origin of the orthopteran auditory system. *J. Neurobiol. 21: 592-610* **SO tymp CO subg homol**

MEIER T REICHERT H 1991 A neuron-specific recognition molecule selectively labels growth cones and developing synapses of individual identified neurons. *In: Elsner N Penzlin H (eds.), Synapse-transmission-modulation: Proc. 19th Göttingen Neurobiol. Conf. Stuttgart, New York: Thieme Verlag. Abstract 493* **devel DN**

MEIER T REICHERT H 1991 Serially homologous development of the peripheral nervous system in the grasshopper. *In: Elsner N Penzlin H (eds.), Synapse-transmission-modulation: Proc. 19th Göttingen Neurobiol. Conf. Stuttgart, New York: Thieme Verlag. Abstract 494* **anat devel mpt leg hom9ol**

MEIER T REICHERT H 1991 Serially homologous development of the peripheral nervous system in the mouthparts of the grasshopper. *J. comp. Neurol. 305: 201-214* **devel mpt leg homol**

MEIER T WOLF H VON HELVERSEN O 1987 "Auditory window" for responsiveness during stridulation of grasshopppers as recorded from one of their auditory interneurons. *In: Elsner N Creutzfeld O (eds.), New frontiers in brain research: Proc. 15th Göttingen Neurobiol. Conf. Stuttgart, New York: Thieme Verlag. Abstract 85* **strid aud IN**

MEINERTZHAGEN IA 1976 The organization of perpendicular fibre pathways in the insect optic lobe. *Phil. Trans. R. Soc. Lond. (Biol.) 274: 555-596* **opt ret anat**

MELIS G 1961 Ricerche sulla fine struttura dei gangli dell'apparato nervoso stomatogastrico negli Ortotteri. *Redia 46: 183-198* **anat SGG gut fineStruc**

MENDAS MV 1948 Histology of the corpora allata of *Melanoplus differentialis. Biol. Bull. Woods Hole 94: 194-207* **retroc CAll anat**

MENGUELLE J FUZEAU-BRAESCH S PAPIN C 1985 The influence of glutathione on the resistance to lindane of the migratory locust, *Locusta migratoria cinerascens. Comp. Biochem. Physiol. C 80: 401-405* **tox**

MEYER V ELSNER N 1989 Combined measurements of tracheal pressure and of respiratory and stridulatory motor patterns in freely moving grasshoppers. *In: Elsner N Singer W (eds.), Dynamics and plasticity in neuronal systems: Proc. 17th. Göttingen Neurobiology Conference. Stuttgart, New York: Thieme Verlag. Abstract 147* **strid resp trach**

MIALL RC 1989 A systems analysis of the role of head position in locust flight. *In: Erber J Menzel R Pflüger H-J Todt D (eds.), Neural mechanisms of behavior: Proc. 2nd Internat. Congr. Neuroethology. Stuttgart & New York: Thieme Verlag. Abstract 2* **flight head neck propr vis**

MIALL RC 1990 Visual control of steering in locust flight: the effects of head movement on responses to roll stimuli. *J. comp. Physiol. A 166: 735-744* **flight head neck propr vis**

MICHEL K PETERSEN M 1982 Development of the tympanal organ in larvae of the migratory locust (*Locusta migratoria*). *Cell Tiss. Res. 222: 667-676* **aud SO tymp devel**

MICHEL R 1973 Rôle des photorécepteurs dans les réactions de vol du criquet pelerin *Schistocerca gregaria* Forskål. *Behaviour 45: 256-263* **oc vis flight OM**

MICHEL R ALBRECHT FO 1978 Comportement de vol et facteurs climatiques chez les acridiens migrateurs. *Behaviour 67: 208-216* **flight temp behav**

MICHEL R BERNARD A 1973 Influence de la pars intercerebralis sur l'induction au vol soutenu chez le criquet pélerin, *Schistocerca gregaria. Acrida 2: 139-150* **flight brain behav**

MICHEL R WEGENER G 1982 Metabolic effects of anoxia in the central nervous system of the locust (*Locusta migratoria*). *Hoppe-Seyler's Z. Physiol. Chem. 363: 1310* **metab**

MICHELSEN A 1966 Pitch discrimination in the locust ear: observations on single sense cells. *J. Insect Physiol. 12: 1119-1131* **aud tymp SO**

MICHELSEN A 1968 Frequency discrimination in the locust ear by means of four groups of receptor cells. *Nature 220: 585-586* **tymp aud SO**

MICHELSEN A 1971 The physiology of the locust ear. I. Frequency sensitivity of single cells in the isolated ear. *Z. vergl. Physiol. 71: 49-62* **aud tymp SO**

MICHELSEN A 1971 The physiology of the locust ear. II. Frequency discrimination based on resonances in the tympanum. *Z. vergl. Physiol. 71: 63-101* **aud tymp SO**

MICHELSEN A 1971 The physiology of the locust ear. III. Acoustical properties of the intact ear. *Z. vergl. Physiol. 71: 102-128* **aud tymp SO**

MICHELSEN A 1973 The mechanics of the locust ear: an invertebrate frequency analyzer. *In: Møller A (ed.): Mechanisms in hearing. London: Academic Press. 911-934* **tymp aud SO**

MICHELSEN A HEDWIG B ELSNER N 1989 Combined biophysical and neurophysiological studies on the interference of sound reception and motor activity in locusts. *In: Elsner N Singer W (eds.), Dynamics and plasticity in neuronal systems: Proc. 17th. Göttingen Neurobiology Conference. Stuttgart, New York: Thieme Verlag. Abstract 149* **strid aud thor SO tymp**

MICHELSEN A LARSEN ON SURLYKKE A 1985 Auditory processing of temporal cues in insect songs: frequency domain or time domain? *In: Michelsen A (ed.): Time resolution in auditory systems. Berlin, Heidelberg, New York: Springer Verlag. 3-27* **aud strid patRec**

MICHELSEN A HEDWIG B ELSNER N 1990 Biophysical and neurophysiological effects of respiration on sound reception in the migratory locust *Locusta migratoria. In: Gribakin FG Wiese K Popov AV (eds.), Sensory systems and communication in arthropods. Basel, Boston, Berlin: Birkhäuser Verlag. 199-203* **aud resp**

MILES CI 1985 Temperature compensation in the nervous system of the grasshopper, *Schistocerca americana. Soc. Neurosci. Abstr. 11, 163* **temp IN DN**

MILES CI 1987 Cooling sensitive mechanoreceptors and temperature compensation in the grasshopper *Schistocerca americana. Soc. Neurosci. Abstr. 13, 140* **temp SO DN**

MILLER BA USHERWOOD PNR 1988 Potassium channels in cultured locust muscle. *Pest. Sci. 24: 85-87* **channel cellCult musc**

MILLER LA 1974 Directional hearing in the locust *Schistocerca gregaria. Soc. Neurosci. Abstr. 4: 339* **aud SO tymp**

MILLER LA 1977 Directional hearing in the locust, *Schistocerca gregaria* (Forskål) (Acrididae, Orthoptera). *J. comp. Physiol. A 119: 85-98* **aud SO tymp**

MILLER PL 1960 Respiration in the desert locust. I. The control of ventilation. *J. exp. Biol. 37: 224-236* **resp MN**

MILLER PL 1960 Respiration in the desert locust. II. The control of the spiracles. *J. exp. Biol. 37: 237-263* **resp MN spir**

MILLER PL 1960 Respiration in the desert locust. III. Ventilation and the spiracles during flight. *J. exp. Biol. 37: 264-278* **resp spir flight MN**

MILLER PL 1965 The central nervous control of respiratory movements. *In: Treherne JE Beament JWL (eds.), The physiology of the insect central nervous system. New York, London: Academic Press. 141-155* **MN spir resp rev**

MILLER PL 1967 The derivation of the motor command to the spiracles of the locust. *J. exp. Biol. 46: 349-371* **resp MN spir**

MILLER PL 1969 Inhibitory nerves to insect spiracles. *Nature 221: 171-3* **resp MN inhib spir**

MILLER R LEAF G USHERWOOD PNR 1973 Blood glutamate in arthropods. *Comp. Biochem. Physiol. A 44: 991-996* **TS glu blood**

MILLER T 1974 Insect visceral muscle. *In: Usherwood PNR (ed.), Insect muscle. New York, London: Academic Press. 545-606* **gut musc**

MILLER TA BENEDECZKY I S-RÓZSA K 1979 Ultrastructure of the muscles of the dorsal diaphragm in *Locusta migratoria*. *Cell Tiss. Res. 203: 93-105* **musc DD fineStruc**

MILLER TA S-RÓZSA K 1981 Control of the alary muscles of locust dorsal diaphragm. *Physiol. Entomol. 6: 51-59* **heart musc DD**

MISRA SD 1945 Studies on the somatic musculature of the desert locust *Schistocerca gregaria* (Forskål, phase gregaria). I. The head. *Indian J. Entomol. 7: 103-38* **anat musc head**

MISRA SD 1947 Studies on the somatic musculature of the desert locust *Schistocerca gregaria* (Forskål). II. The neck and prothorax. *Indian J. Entomol. 8: 1-29* **anat musc thor neck**

MISRA SD 1947 Studies on the somatic musculature of the desert locust *Schistocerca gregaria* (Forskål, phase gregaria). III. The pterothorax. *Indian J. Entomol. 9: 19-72* **anat musc thor flight**

MIZISIN AP JOSEPHSON RK 1987 Mechanical power output of locust flight muscle. *J. comp. Physiol. A 160: 413-419* **musc flight**

MIZISIN AP READY NE 1986 Growth and development of flight muscle in the locust (*Schistocerca nitens* Thunberg). *J. exp. Zool. 237: 45-55* **musc flight devel**

MOBBS PG 1976 Development of the locust ocellus. *Nature 264: 269-271* **oc devel IN SO**

MOBBS PG 1979 Development of the dorsal ocelli of the desert locust *Schistocerca gregaria* Forsk. (Orthoptera: Acrididae). *Int. J. Insect Morphol. Embryol. 8: 237-255* **oc devel IN SO**

MÖHL B 1979 High frequency discharge of the locust wing hinge stretch receptor during flight. *Naturwiss. 66: 158-159* **flight SR propr**

MÖHL B 1985 Sensory aspects of flight pattern generation in the locust. *In: Gewecke M Wendler G (eds.), Insect locomotion. Hamburg & Berlin: Paul Parey Verlag. 139-148* **flight propr**

MÖHL B 1985 The role of proprioception in locust flight control. I. Asymmetry and coupling within the time pattern of motor units. *J. comp. Physiol. A 156: 93-101* **flight MN**

MÖHL B 1985 The role of proprioception in locust flight control. II. Information signalled by forewing stretch receptors during flight. *J. comp. Physiol. A 156: 103-116* **flight MN SR propr**

MÖHL B 1985 The role of proprioception in locust flight control. III. The influence of afferent stimulation of the stretch receptor nerve. *J. comp. Physiol. A 156: 281-291* **flight MN SR propr**

MOHL B 1988 Short-term learning during flight steering in the locust. *In: Elsner N Barth FG (eds.) Sense organs: interfaces between environment and behaviour: Proc.16th. Göttingen Neurobiol. Conf. Stuttgart, New York: Thieme Verlag. Abstract 125* **flight MN thor SR propr learn**

MÖHL B 1988 Short-term learning during flight control in *Locusta migratoria*. *J. comp. Physiol. A 163: 803-812* **flight MN thor SR propr learn**

MÖHL B 1989 Sensorimotor plasticity in the flight system of the locust. *In: Elsner N Singer W (eds.), Dynamics and plasticity in neuronal systems: Proc. 17th. Göttingen Neurobiology Conference. Stuttgart, New York: Thieme Verlag. Abstract 45* **flight propr labil**

MÖHL B 1989 Sense organs and the control of flight. *In: Goldsworthy GJ Wheeler CH (eds.), Insect flight. Boca Raton: CRC Press Inc. 75-97* **flight SO propr rev**

MÖHL B 1989 "Biological noise" and plasticity of sensorimotor pathways in the locust flight system. *J. comp. Physiol. A 166: 75-82* **flight labil MN**

MÖHL B 1989 Function-oriented plasticity of a sensorimotor pathway in the locust flight system. *Naturwiss. 76: 130* **flight propr labil**

MÖHL B 1991 Simulation of a model for motor learning in locust flight control. *In: Elsner N Penzlin H (eds.), Synapse-transmission-modulation: Proc. 19th Göttingen Neurobiol. Conf. Stuttgart, New York: Thieme Verlag. Abstract 87* **flight learn propr**

MÖHL B 1991 Motor learning in the locust flight system requires proprioceptive feedback. *Naturwiss. 78: 84* **flight propr SO learn SR**

MÖHL B BACON J 1983 The tritocerebral commissure giant (TCG) wind-sensitive interneuron in the locust. II. Directional sensitivity and role in flight stabilisation. *J. comp. Physiol. A 150: 453-465* **flight DN**

MÖHL B NACHTIGALL W 1978 Proprioceptive input on the locust flight motor revealed by muscle stimulation. *J. comp. Physiol. A 128: 57-65* **flight propr**

MÖHL B ZARNACK W 1975 Flugsteuerung der Wanderheuschrecke durch Verschiebung der Muskelaktivität. *Naturwiss. 62: 414-442* **flight MN**

MÖHL B ZARNACK W 1977 Activity of the direct downstroke flight muscles of *Locusta migratoria* (L.) during steering behaviour in flight. II. Dynamics of the time shift and changes in the burst length. *J. comp. Physiol. A 118: 235-247* **flight MN**

MOORHOUSE JE FOSBROOKE IHM LUDLOW AR 1987 Stopping a walking locust with sound: an analysis of variation in behavioural threshold. *Exp. Biol. 46: 193-201* **aud walk behav**

MOORHOUSE JE LUDLOW AR FOSBROOKE IHM 1990 The control of walking behaviour in locusts: interactions between competing reflex systems. *Anim. Behav. 39: 613-619* **walk aud behav**

MORAN DT ROWELEY JC VARELA FG 1975 Ultrastructure of the grasshopper proximal femoral chorodotonal organ. *Cell Tiss. Res. 161: 445-457* **SO propr FCO leg fineStruc**

MÖRCHEN A 1980 Spike count and response latency. Two basic parameters encoding sound direction in the CNS of insects. *Naturwiss. 67: 469-470* **aud SO**

MÖRCHEN A RHEINLÄNDER J SCHWARTZKOPFF J 1978 Latency shift in insect auditory nerve fibres: a neuronal time cue of sound direction. *Naturwiss. 65: 656* **aud SO**

MORDUE (LUNTZ) AJ 1979 The role of the maxillary and labial palps in the feeding of *Schistocerca gregaria. Entomol. exp. appl. 25: 279-288* **feed mpt ant SO chem**

MORDUE W 1969 Hormonal control of malpighian tube and rectal function in the desert locust, *Schistocerca gregaria. J. Insect Physiol. 15: 273-285* **pep retroc NSC brain DH**

MORDUE W 1970 Evidence for the existence of diuretic and antidiuretic hormones in locusts. *J. Endocrinol. 46: 119-120* **pep retroc NSC brain DH ADH**

MORDUE W GOLDSWORTHY GJ 1969 The physiological effects of corpus cardiacum extracts in locust. *Gen. comp. Endocrinol. 12: 360-370* **pep retroc CC AKH**

MOREAU R GOURDOUX L BEN KHAY A DUTRIEU J 1984 Hormones des corpora cardiaca et glucagon: effets sur les glucides hémolymphatiques du criquet migrateur nourri et jeûneur. *J. Physiol. (Paris) 80: 14A-15A* **retroc CC pep metab**

MOREAU R GOURDOUX L DUTRIEU J BEN KHAY A 1984 Hemolymph trehalose and carbohydrates in starved male adult *Locusta migratoria*: possibility of endocrine modification. *Comp. Biochem. Physiol. A 78: 481-485* **metab hormX**

MOREAU R GOURDOUX L GIRARDIE J 1988 Neuroparsin: a new energetic neurohormone in the african locust. *Archs. Insect Biochem. Physiol. 8: 135-145* **pep neuropars**

MORETEAU B CHAMINADE N 1983 Effets de quatre insecticides de contact (lindane, feutherin, baygon, deltamethamine) sur le glycérinie et le tréhalosemie au cours du dernière stade larvaire de *Locusta migratoria* (Orthoptère, Acridien). *Ann. Soc. entomol. Fr. N.S. 19: 433-439* **metab tox**

MORETEAU B CHAMINADE N 1988 Effets de l'intoxification par le lindane sur les concentrations en dopamine et 5-hydroxytryptamine dans le cerveau de *Locusta migratoria* (Orthop.; Acrididae) *Ann. Soc. entomol. Fr. N.S. 24: 103-109* **brain TS 5HT dopa tox**

MORETEAU B CHAMINADE N 1990 The effects of lindane poisoning on N-acetyldopamine and N-acetyl 5-hydroxytryptamine concentrations in the brain of *Locusta migratoria* L. (Orthop., Acrididae). *Ecotoxicol. Environ. Safety 20: 115-120* **brain TS 5HT dopa tox**

MORETEAU B JANOTS D 1978 Effets de la réserpine sur le développement d'*Oedipoda coerulescens* L. (Acridien, Orthoptère). *Experientia 34: 1098-1099* **devel tox**

MORETEAU B RAMADE F 1979 Lésions ultrastructurales des corpora cardiaca d'un insecte *Locusta migratoria cinerascens* (Orthoptère) intoxiqué par le lindane. *Zool. Jb. (Physiol.) 83: 340-349* **retroc CC tox fineStruc**

MORETEAU B RAMADE F 1980 Effets de l'intoxication par le lindane sur les corpora allata de *Locusta migratoria* (Orthoptère). *Experientia 36: 1233-1235* **retroc CAll tox**

MORETEAU-LEVITA B 1972 Rôle des lobes optiques dans la réalisation d'homochromie d'*Oedipoda coerulescens* L. (Acridien, Orthoptère). *C. R. hebd. Seanc. Acad. Sci. Paris Ser D 274: 2690* **opt col**

MORETEAU-LEVITA B 1972 Rôle de la pars intercerebralis dans la réalisation d'homochromie d'*Oedipoda coerulescens* L. (Acridien, Orthoptère). *C. R. hebd. Seanc. Acad. Sci. Paris Ser D 274: 3277* **brain col NSC**

MORETEAU-LEVITA B 1972 Rôle des corpora cardiaca dans la pigmentation d'*Oedipoda coerulescens* L. (Acridien, Orthoptère). *C. R. hebd. Seanc. Acad. Sci. Paris Ser D 275: 2699* **retroc col CC**

MORGAN PJ SIEGERT KJ MORDUE W 1987 Preliminary characterization of locust diuretic peptide (DP-1) and another corpus cardiacum peptide (LCCP). *Insect Biochem. 17: 383-388* **pep DH LCCP**

MORTON DB EVANS PD 1983 Octopamine distribution in solitarious and gregarious forms of the locust, *Schistocerca americana gregaria. Insect Biochem. 13. 177-183* **TS oct phase**

MORTON DB EVANS PD 1984 Octopamine release from an identified neurone in the locust. *J. exp. Biol. 113: 269-288* **TS Oct DUM**

MOULINS M GIRARDIE A GIRARDIE J 1974 Manipulation of sexual physiology by brain stimulation in insects. *Nature 250: 339-340* **brain-Stim NSC**

MOUSSA TA 1961 Cytological studies on the association neurones of insects. *Ann. Zool. Agra 3: 161-171* **anat IN brain**

MOUSSA TA BANHAWY M 1960 On a possible relation between vitamin C and the Golgi dictysomes of insect neurones. *J. R. micr. Soc. Lond 78: 111-113* **anat brain metab**

MOUSSA TA BANHAWY M 1961 The cytochemistry of insect neurones. *Ann. Zool. Agra 3: 149-159* **anat brain**

MÜCKE A 1989 Mechanoreceptors of the legs of locusts and their central projection. *In: Erber J Menzel R Pflüger H-J Todt D (eds.), Neural mechanisms of behavior: Proc. 2nd Internat. Congr. Neuroethology. Stuttgart & New York: Thieme Verlag. Abstract 18a* **leg anat SO mech thor**

MÜCKE A 1991 Innervation pattern and sensory supply of the midleg of *Schistocerca gregaria* (Insecta, Orthoptera). *Zoomorphol. 110: 175-187* **SO mech leg anat innerv**

MÜCKE A BICKMEYER U 1989 Morphology and physiology of vibrosensitive receptors and vibratory interneurons in locusts. *In: Elsner N Singer W (eds.), Dynamics and plasticity in neuronal systems: Proc. 17th.*

Göttingen Neurobiology Conference. Stuttgart, New York: Thieme Verlag. Abstract 135 **subg SO propr FCO IN thor vibr**

MÜCKE A LAKES R 1988 Sensory nervous system of intact and regenerated legs of locusts. *In: Elsner N Barth FG (eds.) Sense organs: interfaces between environment and behaviour: Proc.16th. Göttingen Neurobiol. Conf. Stuttgart, New York: Thieme Verlag. Abstract 34* **leg regen SO**

MÜCKE A LAKES R 1989 Regeneration of the midlegs of locusts and central projection of regenerated mechanoreceptors. *In: Elsner N Singer W (eds.), Dynamics and plasticity in neuronal systems: Proc. 17th. Göttingen Neurobiology Conference. Stuttgart, New York: Thieme Verlag. Abstract 136* **regen leg SO**

MUELLER NS 1963 An experimental analysis of molting in embryos of *Melanoplus differentialis. Dev. Biol. 8: 222-240* **tech saline devel moult**

MÜLLER A BÄSSLER D RATHMAYER W 1991 Correlation between inhibitory innervation and muscle fibre heterogeneity in the locust. *In: Elsner N Penzlin H (eds.), Synapse-transmission-modulation: Proc. 19th Göttingen Neurobiol. Conf. Stuttgart, New York: Thieme Verlag. Abstract 52* **musc inhib mech thor innerv**

MÜLLER A WOLF H GALLER S RATHMAYER W 1990 Matching of mechanical responses with neuromuscular properties in insect muscle fibers. *In: Elsner N Roth G (eds.) Brain - Perception - Cognition: Proc. 18th Göttingen Neurobiol. Conf. Stuttgart, New York: Thieme Verlag. Abstract 43* **musc leg NM**

MURPHY VF WANN KT 1985 γ-aminobutyric acid receptor agonist and antagonist interactions in locust extensor tibialis muscle fibres. *Pest. Sci. 16: 537-538* **GABA TS leg musc recept**

MURPHY VF WANN KT 1988 The action of GABA receptor agonists and antagonists on muscle membrane conductance in *Schistocerca gregaria. Brit. J. Pharmacol. 95: 713-722* **TS musc GABA leg recept**

MURRAY MJ 1968 Fibre groups in the auditory nerve of the locust. *Nature 218: 95-96* **aud tymp SO**

MUSKÓ NAGY IZS DEÁK G 1973 Fluorescence microscopy and microspectro-fluorimetry of the monoamines in the brain of *Locusta migratoria migratorioides* R. & F. (Insecta, Orthoptera) with special regard to the protocerebrum. *Annal. Biol. Tihany 40: 85-94* **TS brain**

MUTWALLY HMA 1990 The structure, innervation and function of locust foregut visceral muscle. *Dissertation, University of Lancaster, UK* **innerv gut musc**

MYERS CM BALL EE 1988 Comparative development of the extensor and flexor tibiae muscles in the legs of the locust, *Locusta migratoria. Development 101: 351-361* **devel musc leg**

MYERS CM EVANS PD 1985 An FMRFamide antiserum differentiates between populations of antigens in the ventral nervous system of the locust *Schistocerca gregaria. Cell Tiss. Res. 242: 109-114* **pep FMRF BPP thor abd suboes medN**

MYERS CM EVANS PD 1985 The distribution of bovine pancreatic polypeptide/FMRFamide-like immunoreactivity in the ventral nervous system of the locust. *J. comp. Neurol. 234: 1-16* **pep FMRF BPP thor abd suboes medN**

MYERS CM EVANS PD 1987 An FMRFamide antiserum differentiates between populations of antigens in the brain and retrocerebral complex of the locust *Schistocerca gregaria. Cell Tiss. Res. 250: 93-99* **pep FMRF brain retroc**

MYERS CM EVANS PD 1988 Peripheral neurosecretory cells on the thoracic median nerves of the locust *Schistocerca gregaria. J. Morphol. 195: 45-58* **NSC thor pep anat**

MYERS CM WHITINGTON PM BALL EE 1990 Embryonic development of the innervation of the locust extensor tibiae muscle by identified neurons: formation and elimination of inappropriate axon branches. *Dev. Biol. 137: 194-206* **devel MN leg innerv**

MYERS PZ BASTIANI MJ 1990 Stereotypy and variability in the pattern of the coupling between neurons in the grasshopper. *Soc. Neurosci. Abstr. 16: 811* **devel elect**

NACHTIGALL W 1981 Der Vorderflügel grosser Heuschrecken als Luftkrafterzeuger. I. Modellmessungen zur aerodynamischen Wirkung unterschiedlicher Flügelprofile. *J. comp. Physiol. A 142: 127-134* **flight aero wing**

NAGAYAMA T 1989 Morphology of a new population of spiking local interneurones in the locust metathoracic ganglion. *J. comp. Neurol. 283: 189-211* **thor IN anat local**

NAGAYAMA T 1990 The organization of receptive fields of an antero-medial group of spiking local interneurones in the locust with exteroceptive inputs from the legs. *J. comp. Physiol. A 166: 471-476* **IN thor propr walk leg anat local**

NAGAYAMA T BURROWS M 1989 Functional role of spiking local interneurones of the antero-medial group in the control of the leg movements of the locust. *In: Erber J Menzel R Pflüger H-J Todt D (eds.), Neural* *mechanisms of behavior: Proc. 2nd Internat. Congr. Neuroethology. Stuttgart & New York: Thieme Verlag. Abstract 19* **IN thor propr walk leg anat local**

NAGAYAMA T BURROWS M 1990 Input and output connections of an anteromedial group of spiking local interneurons in the metathoracic ganglion of the locust. *J. Neurosci. 10: 785-794* **IN thor SO mech leg MN local**

NAKANISHI K GOODNOW R KONNO K NIWA M BUKOWNIK R KALLIMOPOULOS TA USHERWOOD PNR ELDEFRAWI AT ELDEFRAWI ME 1990 Philanthotoxin-433 (PhTX-433), a non-competitive glutamate receptor inhibitor. *Pure & appl. Chem. 62: 1223-1230* **TS glu tox recept**

NÄSSEL DR 1988 Neuroactive substances in the insect CNS. *In: Ali MA (ed.) Nervous systems in invertebrates. New York: Plenum Press. 171-212* **5HT oct proct pep GABA TS brain opt**

NÄSSEL DR KLEMM N 1983 Serotonin-like immunoreactivity in the optic lobes of three insect species. *Cell Tiss. Res. 232: 129-140* **5HT TS opt anat**

NAYAR KK 1954 The structure of the corpus cardiacum of *Locusta migratoria*. *Quart. J. micr. Sci. 95: 245-250* **anat retroc CC**

NEL RI 1929 Studies on the development of the genitalia and the genital ducts in insects. I. Females of Orthoptera, Dermaptera. *Quart. J. micr. Sci. 73: 25-85* **genit anat**

NEMCSOK J VIG E HALASY K FEKETE E BENEDECZKY I 1986 Localization and quantitative distribution of biogenic monoamines in the intestinal tract of locust, snail and carp. *Neurochem. Int. 9: 505-510* **gut TS**

NESBITT HHJ 1941 A comparative morphological study of the nervous system of the Orthoptera and related orders. *Ann. entomol. Soc. Am. 34: 51-81* **anat brain**

NEUMANN L 1985 Experiments on tegula function for flight coordination in the locust. *In: Gewecke M Wendler G (eds.), Insect Locomotion. Hamburg & Berlin: Paul Parey Verlag. 149-156* **flight propr SO teg**

NEUMANN L MÖHL B NACHTIGALL W 1982 Quick phase-specific influence of the tegula on the locust flight motor. *Naturwiss. 69: 393-394* **flight teg SO propr**

NEUMANN R LEES G BEADLE DJ BENSON JA 1987 Responses to GABA and other neurotransmitters in insect central neuronal somata in vitro. *In: Hollingworth RM Green MB (eds.), Sites of action for neurotoxic* *pesticides. Washington DC: American Chemical Society. 25-43* **cellCult TS GABA**

NEVILLE AC 1963 Motor unit distribution in the dorsal longitudinal flight muscles in locusts. *J. exp. Biol. 40: 123-136* **flight MN anat musc**

NEVILLE AC WEIS-FOGH T 1963 The effect of temperature on locust flight muscle. *J. exp. Biol. 40: 111-122* **flight musc temp**

NEWLAND PL 1989 Morphology and physiology of ascending interneurones in the metathoracic ganglion of the locust. *In: Erber J Menzel R Pflüger H-J Todt D (eds.), Neural mechanisms of behavior: Proc. 2nd Internat. Congr. Neuroethology. Stuttgart & New York: Thieme Verlag. Abstract 18* **IN thor propr walk leg anat suboes brain**

NEWLAND PL 1990 Morphology of a population of mechanosensory ascending interneurones in the metathoracic ganglion of the locust. *J. comp. Neurol. 199: 242-260* **IN thor propr walk leg anat suboes brain**

NEWLAND PL 1991 Physiological properties of afferents from tactile hairs on the hindlegs of the locust. *J. exp. Biol.155: 487-503* **SO leg mech**

NEWLAND PL 1991 Morphology and somatotopic organization of the central projections of afferents from tactile hairs on the hind leg of the locust. *J. comp. Neurol. 312: 493-508* **SO leg mech anat**

NEWLAND PL BURROWS M 1991 Spiking local neurons: correlation between interneurone morphology and tactile afferent projections in the locust. *Soc. Neurosci. Abstr. 17: 638* **thor IN local anat leg SO mech**

NORTHROP RB 1975 Information processing in the insect compound eye. *In: Horridge GA (ed.), The compound eye and vision of insects. Oxford: Clarendon Press. 378-409* **vis opt rev**

NORTHROP RB 1976 A model for neural signal-to-noise ratio improvement in the insect visual system with implications for anomalous resolution. *Biol. Cybern. 17: 221-235* **DN vis**

NORTHROP RB GUIGON EF 1970 Information processing in the optic lobes of the lubber grasshopper. *J. Insect Physiol. 16: 691-713* **opt vis rom**

NOWEL MS SHELTON PMJ 1981 A Golgi-electron microscopical study of the structure and development of the lamina ganglionaris of the locust optic lobe. *Cell Tiss. Res. 216: 377-401* **opt anat devel fineStruc**

NUTTING WL 1951 A comparative anatomical study of the heart and accessory structures of the orthopteroid insects. *J. Morphol. 89: 784-5* **heart anat**

O'CONNOR TP DUERR JS BENTLEY D 1990 Pioneer growth cone steering decisions mediated by single filopodial contacts in situ. *J. Neurosci. 10: 3935-3946* **devel**

O'CONNOR TP GORODEZKY L TOROIAN-RAYMOND A BENTLEY D 1990 Circumferential guidance cues for pioneer growth cone migration in grasshopper embryonic limb buds. *Soc. Neurosci. Abstr. 16: 625* **devel leg**

O'DELL DA WATKINS BL 1988 The development of GABA-like immunoreactivity in the thoracic ganglia of the locust *Schistocerca gregaria. Cell Tiss. Res. 254: 635-646* **TS GABA devel thor**

O'SHEA M 1971 The antennal cleaning reflex in the desert locust *Schistocerca gregaria* (Forskål). *In: Hemming CF Taylor THC (eds.), Proc. Int. Study Conference on current and future problems of Acridology, 1970. London, Foreign Office. 55-59* **groom behav MN thor ant leg**

O'SHEA M 1972 The antennal cleaning reflex in the locust. *Dissertation, University of Southampton, UK.* **ant groom behav anat leg musc**

O'SHEA M 1975 Two sites of axonal spike initiation in a bimodal interneuron. *Brain Res. 96: 93-98* **DN aud vis IN**

O'SHEA M 1980 Cobalt staining of neurons by microelectrodes. *In: Strausfeld NJ Miller TA (eds.): Experimental entomology: neuroanatomy techniques. Berlin, Heidelberg, New York: Springer Verlag. 341-355* **tech**

O'SHEA M 1985 Are skeletal motoneurons in arthropods peptidergic? *In: Selverston AI (ed.) Model neural networks and behaviour. New York: Plenum Press. 401-413* **musc pep proct mod**

O'SHEA M ADAMS M 1986 Proctolin: from "gut factor" to model neuropeptide. *Adv. Insect Physiol. 19: 1-28* **pep proct mod musc**

O'SHEA M EVANS PD 1979 Potentiation of neuromuscular transmission by an octopaminergic neurone in the locust. *J. exp. Biol. 79: 169-190* **NM TS oct DUM mod**

O'SHEA M HEKIMI S SCHULZ-AELLEN MF 1990 Locust AKH: molecular biology of biosynthesis. *In: Hagedorn HH Hildebrand JG Kidwell MG Law JH (eds.), Molecular insect science. New York: Plenum Press. 189-197.* **pep AKH**

O'SHEA M HEKIMI S WITTEN J WORDEN MK 1988 Functions of aminergic and peptidergic skeletal motoneurons in insects. *In: Thorndyke MC Goldsworthy GJ (eds.): Soc. Exp. Biol. Seminar Series 33. Cambridge University Press. 159-172* **MN pep proct glu mod TS**

O'SHEA M ROWELL CHF 1975 A spike-transmitting electrical synapse between visual interneurons in the locust movement detector system. *J. comp. Physiol. A 97: 143-158* **DN vis opt elect**

O'SHEA M ROWELL CHF 1975 Protection from habituation by lateral inhibition. *Nature 254: 53-55* **DN vis opt habit**

O'SHEA M ROWELL CHF 1976 Neuronal basis of a sensory analyser, the acridid movement detector system. II. Response decrement, convergence, and the nature of the excitatory afferents to the LGMD. *J. exp. Biol. 65: 289-308* **DN vis opt habit**

O'SHEA M ROWELL CHF 1977 Complex neural integration and identified interneurons in the locust brain. *In: Hoyle G (ed.) Identified interneurons and behaviour in arthropods. New York: Plenum Press. 307-328* **DN habit vis opt rev**

O'SHEA M ROWELL CHF WILLIAMS JLD 1974 The anatomy of a locust visual interneurone, the descending contralateral movement detector. *J. exp. Biol. 60: 1-12* **DN vis anat brain thor**

O'SHEA M SCHAFFER M 1985 Neuropeptide function: the invertebrate contribution. *Annu. Rev. Neurosci. 8: 171-198* **pep proct AKH mod musc rev**

O'SHEA M WILLIAMS JLD 1974 Anatomy and output connections of the lobular giant movement detector neuron (LGMD) of the locust. *J. comp. Physiol. 91: 257-266* **vis opt IN anat brain DN**

O'SHEA M WITTEN J SCHAFFER M 1984 Isolation and characterization of two myoactive neuropeptides: further evidence of an invertebrate peptide family. *J. Neurosci. 4: 521-529* **pep leg musc rhythm**

ODHIAMBO TR 1970 The architecture of the accessory reproductive glands of the male desert locust. III. Components of the muscular wall. *Tissue & Cell 2: 233-238* **anat genit musc**

OKELO O 1971 Physiological control of oviposition in the female desert locust *Schistocerca gregaria* (Forsk.). *Can. J. Zool. 49: 969-974* **brain NSC retroc ovip**

OKELO O 1975 Mating and sperm migration in the female *Schistocerca vaga* Scudder. I. The structure and innervation of the spermatheca. *Sci. Biol. J. 1: 144-153* **genit anat MN DUM SO**

OKELO O 1979 Mechanisms of sperm release from the receptaculum seminis of *Schistocerca vaga* Scudder (Orthoptera: Acrididae). *Int. J. Invert. Reprod. Devel. 1: 121-131* **genit anat MN DUM SO**

OLDFIELD AC HUDDART H 1982 Spontaneous activity of foregut and hindgut muscle of the locust, *Locusta migratoria.* I. Normal activity and the effect of KCl depolarization and glutamate. *Comp. Biochem. Physiol. C 73: 297-302* **gut musc glu rhythm TS**

OLTROGGE M HOFMANN E AMMERMÜLLER J 1991 Antigenic determinants in locust (*Locusta migratoria* L.) supraoesophageal ganglion detected with an antiserum against mammalian neurotensin. *In: Elsner N Penzlin H (eds.), Synapse-transmission-modulation: Proc. 19th Göttingen Neurobiol. Conf. Stuttgart, New York: Thieme Verlag. Abstract 350* **brain pep**

ORCHARD I 1982 Octopamine in insects: neurotransmitter, neurohormone, neuromodulator. *Can. J. Zool. 60: 659-699* **oct TS mod**

ORCHARD I 1987 Adipokinetic hormones - an update. *J. Insect Physiol. 33: 451-463* **pep rev AKH retroc oct**

ORCHARD I BELANGER JH LANGE AB 1989 Proctolin: a review with emphasis on insects. *J. Neurobiol. 20: 470-496* **pep proct rev**

ORCHARD I FRIEDEL T LOUGHTON BG 1981 Release of neurosecretory protein from the corpora cardiaca of *Locusta migratoria* induced by high potassium saline and compound action potentials. *J. Insect Physiol. 27: 297-304* **retroc pep CC**

ORCHARD I LANGE AB 1983 Release of identified adipokinetic hormones during flight and following neural stimulation in *Locusta migratoria. J. Insect Physiol. 29: 425-429* **AKH pep flight**

ORCHARD I LANGE AB 1985 Evidence for octopaminergic modulation of an insect visceral muscle. *J. Neurobiol. 16: 171-181* **genit DUM oct musc mod**

ORCHARD I LANGE AB 1986 Neuromuscular transmission in an insect visceral muscle. *J. Neurobiol. 17: 359-372* **NM genit**

ORCHARD I LANGE AB 1987 Octopamine in insects; with special reference to the control of haemolymph lipid and control of visceral muscle in locusts. *Am. Chem. Soc. Symp. Series 356: 136-153* **musc TS oct blood genit mod lip**

ORCHARD I LANGE AB 1987 The release of octopamine and proctolin from an insect visceral muscle: effects of high-potassium saline and neural stimulation. *Brain Res. 413: 251* **musc TS oct pep genit**

ORCHARD I LOUGHTON BG 1980 A hypoglaecemic factor from the corpus cardiacum of *Locusta migratoria. Nature 286: 494-496* **retroc pep CC**

ORCHARD I LOUGHTON BG 1981 The neural control of release of hyperlipaemic hormone from the corpus cardiacum of *Locusta migratoria*. *Comp. Biochem. Physiol. A 68: 25-30* **AKH pep flight retroc CC**

ORCHARD I LOUGHTON BG 1981 Is octopamine a transmitter mediating hormone release in insects? *J. Neurobiol. 12: 143-153.* **oct TS**

ORCHARD I LOUGHTON BG GOLE JWD DOWNER RGH 1983 Synaptic transmission elevates adenosine 3'5'-monophosphate (cyclic AMP) in locust neurosecretory cells. *Brain Res. 258: 152-155* **AKH pep cAMP NSC**

ORCHARD I LOUGHTON BG GOLE JWD DOWNER RGH 1983 Pharmacology of aminergic receptors mediating an elevation in cyclic AMP and release of hormone from locust neurosecretory cells. *Brain Res. 288: 349-353* **retroc cAMP NSC TS recept**

ORCHARD I LOUGHTON BG WEBB RA 1981 Octopamine and short-term hyperlipaemia in the locust. *Gen. comp. Endocrinol. 45: 175-180* **oct TS**

ORCHARD I MARTIN RJ SLOLEY BD DOWNER RGH 1986 The association of 5-hydroxytryptamine, octopamine and dopamine with the intrinsic (glandular) lobe of the corpus cardiacum of *Locusta migratoria*. *Can. J. Zool. 64: 271-274* **retroc TS dopa 5HT oct CC**

ORCHARD I SHIVERS J 1986 Electrical coupling and gap junctions between neurosecretory cells in an insect. *Brain Res. 397: 359-362* **NSC elect fineStruc**

OSBORNE MP 1970 Structure and function of neuromuscular junctions and stretch receptors. *Symp. R. entomol. Soc. Lond. 5: 77-100* **NM SO SR fineStruc abd**

OSBORNE MP FINLAYSON LH RICE MJ 1971 Neurosecretory endings associated with striated muscles in three insects (*Schistocerca, Carausius,* and *Phormia*) and a frog (*Rana*). *Z. Zellforsch. mikr. Anat. 116: 391-404* **NSC musc NM fineStruc**

OSBORNE RH BANNER SE WOOD SJ 1990 The pharmacology of the gut of the desert locust *Schistocerca gregaria* and other insects. *Comp. Biochem. Physiol. C 96: 1-9* **gut pep proct FMRF 5HT oct ACh**

OSORIO D 1986 Directionally sensitive cells in the locust medulla. *J. comp. Physiol. A 159: 841-847* **vis IN**

OSORIO D 1986 Ultraviolet sensitivity and spectral opponency in the locust. *J. exp. Biol. 122: 193-208* **vis IN opt**

OSORIO D 1987 The temporal properties of non-linear, transient cells in the locust medulla. *J. comp. Physiol. A 161: 431-440* **vis IN opt**

OSORIO D 1987 Spectral and temporal properties of sustaining cells in the medulla of the locust. *J. comp. Physiol. A 161: 441-448* **vis IN opt**

OSORIO D 1991 Mechanisms of early visual processing in the medulla of the locust optic lobe: how self-inhibition, spatial-pooling and signal rectification contribute to the properties of transient cells. *Visual Neurosci. 7: 345-355* **vis opt IN**

OTTO C KALMRING K LORENZEN S SIPPEL M 1983 Problems of recordings of nervous activity with glass microelectrodes in the CNS of insects. *Biol. Cybern. 49: 63-67* **tech**

ÖZBAS S 1957 Morphological and histological studies on the corpora allata and cardiaca in Orthoptera. *Comm. Fac. Sci. Ankara C 8: 19-44* **retroc anat CAll CC**

ÖZBAS S 1957 Two kinds of secretions in corpora cardiaca of *Locusta migratoria* (L.) ph. *solitaria* (Orthoptera: Acrididae). *Comm. Fac. Sci. Ankara C 8: 45-47* **retroc pep**

PABST H 1965 Elektrophysiologische Untersuchungen des Streckrezeptors am Flügelgelenk der Wanderheuschrecke *Locusta migratoria. Z. vergl. Physiol. 5: 498-541* **SR propr flight**

PABST H SCHWARZKOPF J 1962 Zur Leistung der Flügelgelenkrezeptoren von *Locusta migratoria. Z. vergl. Physiol. 45: 396-404* **flight SR propr**

PACICCA C REICHERT H 1990 Development of the enteric nervous system of the locust. *In: Elsner N Roth G (eds.), Brain - Perception - Cognition: Proc. 18th Göttingen Neurobiol. Conf., Stuttgart, New York: Thieme. Abstract 370* **devel SGG**

PADGHAM DE 1976 Control of melanization in first instar larvae of *Schistocerca gregaria. J. Insect Physiol. 22:1409-1419* **bursicon pep col**

PADGHAM DE 1976 Bursicon mediated control of tanning in melanizing and non-melanizing larvae of *Schistocerca gregaria. J. Insect Physiol. 22:1447-1452* **bursicon pep col phase**

PAGE CH 1970 Unit responses in the metathoracic ganglion of the flying locust. *Comp. Biochem. Physiol. 37: 565-572* **flight IN thor**

PALKA J 1965 Diffractive limitations on visual acuity and some properties of a visual pathway in the ventral nerve cord of locusts. *Dissertation, University of California at Los Angeles, USA.* **vis DN**

PALKA J 1965 Diffraction and visual acuity in insects. *Science 149: 551-553* **vis DN**

PALKA J 1967 Head movement inhibits locust visual unit's response to target movement. *Am. Zool. 7: 728 (Abstract)* **vis DN**

PALKA J 1967 An inhibitory process influencing visual responses in a fibre of the ventral nerve cord of locusts. *J. Insect Physiol. 13: 235-248* **vis DN**

PALKA J 1972 Moving movement detectors. *Am. Zool. 12: 497-505* **vis DN**

PALKA J PINTER RB 1974 Theoretical and experimental analysis of visual acuity in insects. *In: Horridge GA (ed.), The compound eye and vision of insects. Oxford: Clarendon Press. 321-337* **vis DN**

PANNABECKER T ORCHARD I 1986 Pharmacological properties of octopamine-2 receptors in locust neuroendocrine tissue. *J. Insect Physiol. 32: 909-915* **TS oct NSC retroc CC recept**

PANNABECKER T ORCHARD I 1986 Octopamine and cyclic AMP mediate release of adipokinetic hormone I and II from isolated locust neuroendocrine tissue. *Mol. cell. Endocrinology 48: 153-159* **AKH NSC pep oct retroc CC**

PANNABECKER T ORCHARD I 1987 Regulation of adipokinetic hormone release from locust neuroendocrine tissue: participation of calcium and cyclic AMP. *Brain Res. 423: 13-22* **AKH NSC pep oct cAMP retroc CC**

PANNABECKER T ORCHARD I 1988 Evidence for a Na+/Ca2+ exchange mechanism in the neuroendocrine cells of the locust corpus cardiacum. *Brain Res. 458: 180* **retroc NSC CC**

PANNABECKER T ORCHARD I 1988 Receptors for octopamine in the storage lobe of the locust (*Locusta migratoria*) corpus cardiacum: evidence from studies on cyclic nucleotides. *J. Insect Physiol. 34: 815-820* **retroc oct TS CC cAMP recept**

PANNABECKER T ORCHARD I 1989 Ionic dependence of depolarization-mediated adipokinetic hormone release from the locust corpus cardiacum. *Brain Res. 477: 38* **pep AKH retroc CC**

PANNABECKER T ORCHARD I LANGE AB 1987 Adipokinetic hormones in neuroendocrine tissue of the larval locust: quantification and regulation of in vitro release. *J. Insect Physiol. 33: 575-580* **AKH NSC pep retroc CC**

PANOV AA 1962 [Distribution of neurosecretory cells in the abdominal portion of the neural chain in Orthoptera]. (in Russian). *Dok. Akad. Nauk. SSSR 145: 1409-1412* **NSC abd anat**

PANOV AA 1966 A comparison of the neurosecretory systems in related species of insects. *In: Novak VJA Slama L (eds.), Insect endocrines. (Proc. Int. Symp. Insect Endocrinol., Brno 1966). Prague: Academia. 91-95* **NSC abd anat**

PANOV AA 1978 Neurosecretory cells in the pars intercerebralis of Orthoptera. *Zool. Anz. 201: 49-63* **NSC brain anat**

PANOV AA 1982 On the homologies of the ventral neurosecretory cells of Orthopteran brain (Insecta). *Zool. Anz. 108: 485-492* **NSC brain anat**

PANOV AA 1988 On the evolution of the retrocerebral endocrine complex in Orthoptera (Insecta). *Zool. Anz. 221: 379-385* **NSC retroc anat**

PAPILLON M CASSIER P GIRARDIE J LAFON-CAZAL M 1976 L'influence de la température d'élevage sur l'activité endocrine de *Schistocerca gregaria. Archs. Biol. Bruxelles 87: 103-127* **NSC brain temp**

PARIHAR DR 1972 The embryonic stages of the ik grasshopper, *Poekilocerus pictus* (Acridoidea: Pyrgomorphidae). *J. Anim. Morphol. Physiol. 19: 16-22* **devel pyrg**

PARRY DA 1947 The function of the insect ocellus. *J. exp. Biol. 24: 211-219* **oc DN**

PATEL NH MARIN-BLANCO E COLEMAN KG POOLE SJ ELLIS MC KORNBERG TB GOODMAN CS 1989 Expression of engrailed proteins in arthropods, annelids, and chordates. *Cell 58: 955-968* **devel**

PATLAK JB GRATION KAF USHERWOOD PNR 1979 Single glutamate activated channels in locust muscle. *Nature 278: 643-645* **glu musc channel NM TS**

PATTERSON JA GOODMAN LJ 1974 Intracellular responses of receptor cells and second-order cells in the ocelli of the desert locust *Schistocerca gregaria. J. comp. Physiol. A 95: 237-250* **oc IN ret brain**

PATTERSON JA GOODMAN LJ 1974 Relationship between ocellar units in the ventral nerve cord and ocellar pathways in the brain of *Schistocerca gregaria. J. comp. Physiol. A 95: 251-262* **oc IN DN brain**

PATTERSON JA GOODMAN LJ 1975 Component analysis of the ocellar electroretinogram of the locust *Schistocerca gregaria. J. Insect Physiol. 21: 287-298* **oc**

PAULUS U BRÄUNIG P FISCHER FP 1989 The ultrastructure of the locust satellite nervous system. *In: Elsner N Singer W (eds.), Dynamics and plasticity in neuronal systems: Proc. 17th. Göttingen Neurobiology Conference. Stuttgart, New York: Thieme Verlag. Abstract 52* **suboes NSC TS 5HT fineStruc**

PAVLOVA M 1985 [On the structure of the blood vessels and sympathetic nervous system, especially of Orthoptera.] (in Russian) *Raboty Lab. Zool. Varshavsk. Univ. (1895): 1-95* **anat retroc heart**

PAYNE R HOWARD J 1981 Response of an insect photoreceptor: a simple log-normal model. *Nature 290: 415-417* **ret**

PEARSON KG 1976 Local neurons and local interactions in the nervous systems of invertebrates. *In: Schmitt FO Worden FG (eds.), The neurosciences, fourth study program. Boston, Mass.: M.I.T. Press. 145-160* **IN local**

PEARSON KG 1977 Interneurons in the central nerve cord of insects. *In: Hoyle G (ed.): Identified neurons and behavior in arthropods. New York: Plenum Press. 329-338* **IN thor anat**

PEARSON KG 1980 Burst generation in coordinating interneurons of the ventilatory system of the locust. *J. comp. Physiol. A 137: 305-313* **resp IN thor anat rhythm**

PEARSON KG 1981 Function of sensory input in insect motor systems. *Can. J. Physiol. Pharmacol. 59: 660-666* **SO**

PEARSON KG 1982 Neural circuits for jumping in the locust. *J. Physiol. (Paris) 78: 765-771* **jump IN thor rev**

PEARSON KG 1985 Are there central pattern generators for walking and flight in insects? *In: Barnes WJP Gladden MH (eds.), Feedback and motor control. London, Sydney, Dover, New Hampshire: Croom Helm Ltd. 307-315* **flight walk rhythm**

PEARSON KG 1985 Neuronal circuits for patterning motor activity in invertebrates. *In: Strumwasser F Cohen M (eds.): Comparative neurobiology: modes of communication in the nervous system. New York: John Wiley & Sons Inc. 225-244* **IN flight jump**

PEARSON KG 1987 Central pattern generation: a concept under scrutiny. *In: McLennan H Ledsome JR McIntosh CHS Jones DR (eds.): Advances in physiological research. New York: Plenum Press. 167-186* **rhythm IN**

PEARSON KG BERGMAN SJ 1969 Common inhibitory motoneurones in insects. *J. exp. Biol. 50: 445-472* **MN inhib anat**

PEARSON KG BOYAN GS BASTIANI M GOODMAN CS 1985 Heterogeneous properties of segmentally homologous interneurones in the ventral cord of locusts. *J. comp. Neurol. 223: 133-145* **flight IN homol anat var devel MN flight**

PEARSON KG FRANKLIN R 1984 Characteristics of leg movements and patterns of coordination in locusts walking on rough terrain. *Int. J. Robotics 3: 101-112* **walk behav**

PEARSON KG GOODMAN CS 1979 Correlation of variability of structure with variability of synaptic connection of an identified interneuron in locust. *J. comp. Neurol. 184: 141-165* **DN var MN anat**

PEARSON KG GOODMAN CS 1981 Presynaptic inhibition of transmission from identified interneurons in locust central nervous system. *J. Neurophysiol. 45: 501-515* **DN jump inhib**

PEARSON KG GYNTHER IC HEITLER WJ 1986 Coupling of flight initiation to the jump in locusts. *J. comp. Physiol. A 158: 81-89* **flight jump IN**

PEARSON KG HEDWIG B WOLF H 1989 Are the hindwing chordotonal organs elements of the locust flight pattern generator? *J. exp. Biol. 144: 235-255* **flight SO CO aud anat IN thor rhythm**

PEARSON KG HEITLER WJ STEEVES JD 1980 Triggering of locust jump by multimodal inhibitory interneurons. *J. Neurophysiol. 43: 257-278* **jump IN anat**

PEARSON KG O'SHEA M 1984 Escape behavior in locust. *In: Eaton RC (ed.), The neural basis of startle behavior. New York: Plenum Press. 163-177* **jump behav rev**

PEARSON KG RAMIREZ JM 1990 Influence of input from the forewing stretch receptors on motoneurones in flying locusts. *J. exp. Biol. 151: 317-340* **MN SR flight propr thor**

PEARSON KG REYE DN PARSON DW BICKER G 1985 Flight initiating interneurons in the locust. *J. Neurophysiol. 53: 910-934* **flight IN anat**

PEARSON KG REYE DN ROBERTSON RM 1983 Phase dependent influence of wing stretch receptors on flight rhythms in the locust. *J. Neurophysiol. 49: 1168-1181* **flight SR propr anat**

PEARSON KG ROBERTSON RM 1981 Interneurons coactivating hindleg flexor and extensor motoneurons in the locust. *J. comp. Physiol. A 144: 391-400* **IN thor leg jump MN anat**

PEARSON KG ROBERTSON RM 1987 Structure predicts synaptic function of two classes of interneurons in the thoracic ganglia of *Locusta migratoria. Cell Tiss. Res. 250: 105-114* **flight IN anat**

PEARSON KG ROWELL CHF 1977 Functions of tonic sensory input in insects. *Ann. N. Y. Acad. Sci.: 114-123* **SO**

PEARSON KG WOLF H 1987 Comparison of motor patterns in the intact and deafferented flight system of the locust. I. Electromyographic analysis. *J. comp. Physiol. A 160: 259-268* **flight MN SO**

PEARSON KG WOLF H 1988 Connections of hindwing tegulae with flight neurones in the locust *Locusta migratoria. J. exp. Biol. 135: 381-409* **flight IN MN teg propr SO**

PEARSON KG WOLF H 1989 Timing of forewing elevator activity during flight in the locust. *J. comp. Physiol. A 165: 217-227* **flight MN**

PECE AEC FRENCH AS 1989 Single photon responses in locust photoreceptors: the effects of stimulus location on amplitude and time course. *J. comp. Physiol. A 164: 365-376* **ret saline**

PECE AEC FRENCH AS KORENBERG MJ KUSTER JE 1990 Nonlinear mechanisms for gain adaptation in locust photoreceptors. *Biophys. J. 57: 733* **ret**

PENER MP 1983 Endocrine aspects of phase polymorphism in locusts. *In: Downer RGH Laufer H (eds.), Invertebrate endocrinology, Vol. 1: endocrinology of insects. New York: Alan R. Liss Inc. 379-394* **brain retroc phase**

PENER MP 1985 Hormonal effects on flight and migration. *In: Kerkut GA Gilbert LI (eds.), Comprehensive insect physiology, biochemistry and pharmacology, Vol. 8. London: Pergamon Press. 491-550* **flight**

PENER MP GIRARDIE A JOLY P 1972 Neurosecretory and corpora allata controlled effects on mating behaviour and color change in adult *Locusta migratoria migratorioides* males. *Gen. comp. Endocrinol. 19: 494-508* **NSC retroc CAll col behav**

PERDECK AC 1958 The isolating value of specific song patterns in two sibling species of grasshoppers (*Chorthippus brunneus* Thunb. and *C. biguttulus* L.). *Behaviour 12: 1-75* **strid behav**

PETERS BH 1987 Novel synaptic structures in the central nervous system of the locust *Schistocerca gregaria. Cell Tiss. Res. 249: 341-350* **fineStruc**

PETERS BH ALTMAN JS TYRER NM 1985 Synaptic connections between the hindwing stretch receptor and flight motor neurones in the locust revealed by double cobalt labelling for electron microscopy. *J. comp. Neurol. 254: 34-50* **fineStruc SR MN flight propr tech**

PETERS BH BUTLER SV TYRER NM 1987 Morphology, ultrastructure and synapse distribution of putative serotonergic salivary neurons in the locust. *Neuroscience 23: 705-719* **5HT saliv MN anat fineStruc suboes**

PETERS BH RÖMER H MARQUART V 1986 Spatial segregation of synaptic inputs and outputs in a locust auditory interneurone. *J. comp. Neurol. 254: 34-50* **aud fineStruc IN**

PETERS BH TYRER NM 1987 Electron microscopy of serotonin-immunoreactive neurone branches and terminals in the locust central nervous system. *Neuroscience 23: 333-341* **TS 5HT fineStruc**

PETERS M 1977 Innervation of the ventral diaphragm of the locust (*Locusta migratoria*). *J. exp. Biol. 69: 23-32* **VD innerv**

PETERSEN M KALMRING K COKL A 1982 The auditory system in larvae of the migratory locust. *Physiol. Entomol. 7: 43-54* **aud SO tymp devel anat**

PFAU HK 1976 Ein Muskel zur Feineinstellung des Vorderflügelanstellwinkels bei *Locusta migratoria* (L.). *Verh. dt. zool. Ges. 1976: 246* **flight musc anat**

PFAU HK 1977 Zur Morphologie und Funktion des Vorderflügels und Vorderflügelgelenks von *Locusta migratoria* L. *Fortschr. Zool. 24: 341-345* **flight musc anat**

PFAU HK 1978 Bedeutung des Vorderflügel-Streckrezeptors und -Chordotonalorgans für die Registrierung der geometrischen Flügelanstellung bei *Locusta migratoria. Verh. dt. zool. Ges. 1978: 247* **flight anat SR CO propr SO aero**

PFAU HK 1978 Funktionsanatomische Aspekte des Insektenflugs. *Zool. Jb. (Anat.) 99: 99-108* **flight musc anat**

PFAU HK 1983 Mechanik und sensorische Kontrolle der Flügel-Pronation und -Supination. *In: Nachtigall W (ed.), BIONA-report 1 "Insekten Flug", Akad. Wiss. Mainz. Stuttgart & New York: Gustav Fischer Verlag. 61-77* **flight musc anat SO**

PFAU HK KOCH UT MÖHL B 1989 Temperature dependence and response characteristics of the isolated wing hinge stretch receptor in the locust. *J. comp. Physiol. A 165: 247-252* **flight SR temp SO**

PFAU HK NACHTIGALL W 1981 Der Vorderflügel grosser Heuschrecken als Luftkrafterzeuger. II. Zusammenspiel von Muskeln und Gelenkmechanik bei der Einstellung der Flügelgeometrie. *J. comp. Physiol. A 142: 135-140* **flight anat aero**

PFLÜGER H-J 1980 The function of hair sensilla on the locust's leg: the role of tibial hairs. *J. exp. Biol. 87: 163-175* **SO local mech leg**

PFLÜGER H-J 1980 Central nervous projections of sternal trichoid sensilla in locusts. *Naturwiss. 67: 316* **SO anat thor windh mech**

PFLÜGER H-J 1984 The large fourth abdominal intersegmental interneuron: a new type of wind-sensitive ventral cord interneuron in locusts. *J. comp. Neurol. 222: 343-357* **IN windh SO mech abd**

PFLÜGER H-J 1988 Wie laufen Tiere? *Ber. Ges. Naturf. Freunde Berlin. (N.F.) 28: 109-120* **SO IN walk rev**

PFLÜGER H-J 1988 Control of local reflexes in insects. *In: Amblard B Berthoz A Clarac F (eds.), Posture and gait: development, adaptation and modulation. Amsterdam: Elsevier Science Publishers BV. 321-331* **SO mech tymp propr anat IN rev**

PFLÜGER H-J BURROWS M 1978 Locusts use the same basic motor patterns in swimming as in jumping and kicking. *J. exp. Biol. 75: 81-93* **jump swim kick**

PFLÜGER H-J BRÄUNIG P HUSTERT R 1981 Distribution and specific central projections of mechanoreceptors in the thorax and proximal leg joints of locusts. II. The external mechanoreceptors, hair plates and tactile hairs. *Cell Tiss. Res. 216: 79-96* **SO anat thor mech**

PFLÜGER H-J BRÄUNIG P HUSTERT R 1988 The organization of mechanosensory neuropiles in locust thoracic ganglia. *Phil. Trans. R. Soc. Lond. (Biol.) 321: 1-26* **SO anat mech thor**

PFLÜGER H-J BURROWS M 1978 How the locust dries itself. *J. exp. Biol. 75: 95-100* **kick**

PFLÜGER H-J BURROWS M 1987 A strand receptor with a central cell body synapses upon spiking local interneurones in the locust. *J. comp. Physiol. A 160: 295-304* **SO local propr IN thor**

PFLÜGER H-J BURROWS M 1988 Positive feedback loops by campaniform sensilla in the locust jump system. *In: Elsner N Barth FG (eds.), Sense organs: interfaces between environment and behaviour: Proc.16th. Göttingen Neurobiol. Conf. Stuttgart, New York: Thieme Verlag. Abstract 19* **SO kick leg thor MN mech**

PFLÜGER H-J BURROWS M 1990 Synaptic connections of different strength between wind-selective hairs and an identified projection interneuron in the locust. *Eur. J. Neurosci. 2: 1040-1050* **windh SO mech IN thor**

PFLÜGER H-J BURROWS M CZJZEK A 1990 The locust prosternal hair system: Neurobiology and development. *In: Elsner N Roth G (eds.) Brain - Perception - Cognition: Proc. 18th Göttingen Neurobiol. Conf. Stuttgart, New York: Thieme Verlag. Abstract 37* **thor windh SO mech IN**

PFLÜGER H-J ELSON R BINKLE U SCHNEIDER H 1986 The central nervous organization of the motor neurones to a steering muscle in locusts. *J. exp. Biol. 120: 403-420* **flight MN**

PFLÜGER H-J KUTSCH W 1987 Pionier- und Wegweiser-Zellen im Nervensystem der Insekten. *Biologie in unserer Zeit 17: 1-8* **devel rev**

PFLÜGER H-J STEVENSON P FERBER M ECKERT M 1990 Neurones with bilateral axons in the locust central nervous system. *Soc. Neurosci. Abstr. 16: 1225* **DUM oct pep FMRF heart**

PFLÜGER H-J TAUTZ J 1982 Air movement sensitive hairs and interneurons in *Locusta migratoria. J. comp. Physiol. A 145: 369-380* **windh SO mech IN thor**

PFLÜGER H-J WATSON AHD 1987 The system of abdominal, neuromodulatory DUM-cells in locusts. *In: Elsner N Creutzfeld O (eds.), New frontiers in brain research: Proc. 15th Göttingen Neurobiol. Conf. Stuttgart, New York: Thieme Verlag. Abstract 246* **DUM genit abd**

PFLÜGER H-J WATSON AHD 1988 Structure and distribution of dorsal unpaired median (DUM) neurones in the abdominal nerve cord of male and female locusts. *J. comp. Neurol. 268: 329-345* **DUM abd anat genit**

PHILLIPS CE 1980 An arthropod muscle innervated by nine excitatory motor neurones. *J. exp. Biol. 88: 249-258* **MN musc anat leg thor innerv**

PHILLIPS CE 1980 Intracellularly injected cobaltous ions accumulate at synaptic densities. *Science 207: 1477-1479* **fineStruc tech saline**

PHILLIPS CE 1981 Organisation of motor neurons to a multiply-innervated insect muscle. *J. Neurobiol. 12: 269-280* **MN leg thor**

PHILLIPS JE MORDUE W MEREDITH J SPRING J 1980 Purification and characteristics of the chloride transport stimulating factor from locust corpora cardiaca: a new peptide. *Can. J. Zool. 58: 1851-1860* **pep retroc CC**

PICHLER F 1953 Beitrag zur Kenntnis der tympanalen Muskulatur der Feldheuschrecken. *Zool. Anz. 15: 12-15* **SO tymp musc anat aud**

PICQUOT M PROUX J 1987 Relationship between excretion of primary urine and haemolymphatic level of diuretic hormone along a circadian cycle in the migratory locust. *Physiol. Entomol. 12: 455-460* **pep blood DH**

PICQUOT M PROUX J 1989 Influence of hydration on the concentration of two arginine-vasopressin-like neuropeptides coexisting in the ventral nerve cord of the migratory locust, *Locusta migratoria. J. Insect Physiol. 35: 571-577* **pep**

PICQUOT M PROUX J 1990 Biosynthesis and degradation of the arginine vasopressin-like insect diuretic hormone, a migratory locust neurohormone. *Regulatory Peptides 34: 139-156* **pep DH**

PIEK T MANTEL P 1977 Myogenic contractions in locust muscle induced by proctolin and by wasp Philanthus triangulum venom. *J. Insect Physiol. 23: 312-325* **NM rhythm musc pep proct**

PILZ J ELSNER N 1988 The neuromuscular activity during the turning reaction as an indicator for directional sound recognition of the grasshopper *Chorthippus biguttulus. In: Elsner N Barth FG (eds.) Sense organs: interfaces between environment and behaviour: Proc.16th. Göttingen Neurobiol. Conf. Stuttgart, New York: Thieme Verlag. Abstract 139* **aud walk thor musc**

PINTER RB 1972 Frequency and time domain properties of retinular cells of the desert locust *Schistocerca gregaria* and the house cricket *Acheta domesticus. J. comp. Physiol. 77: 383-398* **ret gryll**

PINTER RB 1977 Visual discrimination between small objects and large textured backgrounds. *Nature 270: 429-431* **vis DN**

PINTER RB 1979 Inhibition and excitation in the locust DCMD receptive field: spatial frequency, temporal and spatial characteristics. *J. exp. Biol. 80: 191-216* **vis DN**

PINTER RB OLBERG RM ABRAMS TW 1982 Is the locust DCMD a looming detector? *J. exp. Biol. 101: 327-332* **DN vis**

PLOTNIKOVA SI 1967 [The innervation of the gut in *Locusta migratoria* L. (Orthoptera, Acrididae)]. (In Russian). *Entomol. Obozr. 46 : 122-126. See also Entomol. Rev. - Engl. Transl. 6: 69-71* **anat gut innerv**

PLOTNIKOVA SI 1968 The structure of the sympathetic nervous system of insects. *In: Salánki J (ed.), Symp. Neurobiol. Invert., Tihany 1967. New York: Plenum Press. 59-68* **anat medN TS ACh SGG**

PLOTNIKOVA SI 1969 [The structure of the sympathetic centres of the nervous system of insects]. (In Russian). *Tr. Vses. Entomol. Obshchest. 53: 36-59* **anat**

PLOTNIKOVA SI 1969 [Effectory neurons with several axons in the ventral nerve cord of *Locusta migratoria.*] (In Russian) *Zh. evol. Biokhim. Fiziol. 5: 339-341* **anat DUM thor Abd**

PLOTNIKOVA SI 1975 [On sensory neuropil of nerve cord ganglia of *Locusta migratoria* L.] (In Russian) *Proc. Leningrad Soc. Nature Study 77: 5-21* **anat thor**

PLOTNIKOVA SI 1977 [Effectory neurons of thoracic ganglia in *Locusta migratoria* L.] (In Russian) *In: [Nerve cord of Arthropoda and Vermes]. (In Russian). Leningrad: Nauka. 135-167* **MN thor**

PLOTNIKOVA SI 1979 *[Structural organization of the insect nervous system.] (In Russian). Leningrad: Nauka. 119 pp.* **anat**

PLOTNIKOVA SI 1988 [Motor region of the suboesophageal ganglion in the locust and the larval dragonfly]. (In Russian). *Zh. evol. Biokhim. Fiziol. 24: 232-239. See also J. evol. Biochem. Physiol. - Engl. Transl. 24: 172-178* **anat suboes MN odon**

PLOTNIKOVA SI 1989 [Dorsal connective axons of the suboesophageal ganglion in insects]. (In Russian). *Zh. evol. Biokhim. Fiziol. 25: 525-531. See also J. evol. Biochem. Physiol. - Engl. Transl. 25: 375-380* **anat suboes SO brain DN**

PLOTNIKOVA SI 1989 [Sensory neuropil in the insect suboesophageal ganglion]. (In Russian). *Zh. evol. Biokhim. Fiziol. 25: 95-103. See also J. evol. Biochem. Physiol. - Engl. Transl. 25: 74-80* **anat suboes SO**

PLOTNIKOVA SI NEVMYVAKA GA 1980 The methylene blue technique: classic and recent applications to the insect nervous system. *In: Strausfeld NJ Miller TA (eds.), Neuroanatomical techniques. Insect nervous system. Berlin, Heidelberg, New York: Springer Verlag. 429-446* **anat tech**

POCOCK JM NICHOLSON RA OSBORNE MP 1990 Effects of colchicine on central and peripheral nerve terminals of the locust *Schistocerca gregaria. J. Neurocytol. 19: 574-583* **NM fineStruc**

POCOCK JM OSBORNE MP NICHOLSON RA 1990 The effect of in vivo stimulation on the cytology of neuromuscular junctions of locust flight muscles. *J. Neurocytol. 19: 566-573* **NM fineStruc flight**

POND CM 1972 Neuromuscular activity and wing movements at the start of flight of *Periplaneta americana* and *Schistocerca gregaria. J. comp. Physiol. 78: 192-209* **flight MN**

POND CM 1972 The initiation of flight in unrestrained locusts, *Schistocerca gregaria. J. comp. Physiol. 80: 163-178* **flight**

POPOV AV 1965 [Electrophysiological studies of peripheral auditory neurons in the locust.] (In Russian) *J. evol. Biochem. Physiol. 1: 239-250* **aud tymp SO**

POPOV AV 1966 [The structure and function of receptors in the auditory organ of insects]. *In: Samsonova VG (ed.), [The primary processes in receptor elements of sense organs]. Leningrad: Nauka. 144-154* **tymp SO aud**

POPOV AV 1966 The functional organization of the central part of the locust hearing system (Acrididae, Orthoptera). *Proc. 18th. Int Congr. Psychol. Moscow, Symp 1966, Vol 15. 177-187* **aud IN thor**

POPOV AV 1967 [Synaptic transmission at the level of the first synapses in the auditory pathway of the locust, *Locusta migratoria* L.] (in Russian) *In: Kreps EM (ed.), [Evolutionary neurophysiology and neurochemistry]. Leningrad: Nauka. 54-67* **aud tymp SO**

POPOV AV 1967 [Characteristics of central neurons in the auditory system of the locust]. (in Russian). *In: Gersuni GV (ed.), [Problems of physiological acoustics. Vol. 6: Mechanisms of Hearing]. 108-121* **aud IN thor**

POPOV AV 1968 [An electrophysiological investigation of the functional properties of peripheral and central neurons in the locust auditory system.] *Dissertation, Academy NAUK USSR, Leningrad, USSR. 231 pp..* **aud tymp IN SO**

POPOV AV 1968 [Principles of structural and functional organization of the first levels in the tympanal auditory system of different insects] (in Russian). *Proc. 6th. Acoust. Conf., Moscow, 1968. Section I-2: 4-8* **aud tymp fineStruc**

POPOV AV 1969 [Comparative investigation of sound signals of communication and some principles of the hearing system and organization in cicadas and Orthoptera.] (in Russian) *In: Voskrenskaya AK (ed.), [Modern problems of stucture and function of the nervous system of insects]. Tr. Vses. Entomol. Obsch. 53: 182-221* **aud thor strid**

POPOV AV 1971 Synaptic transformation in the auditory system of insects. *In: Sensory processes at the neuronal and behavioural level. Academic Press, New York & London. 301-320* **tymp SO IN aud**

POPOV AV SHUVALOV VF SVETLOGORSKAYA ID MARKOVICH AM 1975 Acoustic behaviour and auditory system in insects. *In: Schwartzkopff J (ed.), Mechanoreception. Abh. Rheinisch-Westphalische Akad. Wissensch. 53: 281-306. Opladen: Westdeutscher Verlag, 281-306* **aud**

POPOV AV SVETLOGORSKAYA ID 1971 [Ultrastructural organization of the locust auditory nerve (to the question of receptor interaction)]. (in Russian) *Zh. evol. Biokhim. Fiziol. 7: 516-521* **aud SO tymp fineStruc**

POTENTE A 1975 Untersuchung zur Morphologie der cercalen Riesenfasern im Bauchmark von *Locusta migratoria. Diplomarbeit, Ruhr-Universität Bochum, Germany.* **IN GF abd anat**

PRATT GE PENER MP 1983 Precocence sensitivity of corpora allata in adult female *Locusta migratoria* after electrocoagulation of the pars intercerebralis neurosecretory cells. *J. Insect Physiol. 29: 33-39* **retroc CAll brain NSC**

PREISS R 1989 Visual control of migratory flight in the desert locust *Schistocerca gregaria. In: Elsner N Singer W (eds.), Dynamics and plasticity in neuronal systems: Proc. 17th. Göttingen Neurobiology Conference. Stuttgart, New York: Thieme Verlag. Abstract 112* **flight OM**

PREISS R 1991 Set point of retinal velocity of ground images in the control of swarming flight in desert locusts. *In: Elsner N Penzlin H (eds.), Synapse-transmission-modulation: Proc. 19th Göttingen Neurobiol. Conf. Stuttgart, New York: Thieme Verlag. Abstract 89* **flight vis behav**

PREISS R GEWECKE M 1988 Visually-induced wind compensation in the migratory flight of the desert locust *Schistocerca gregaria. In: Elsner N Barth FG (eds.) Sense organs: interfaces between environment and behaviour: Proc.16th. Göttingen Neurobiol. Conf. Stuttgart, New York: Thieme Verlag p. 38* **flight vis OM**

PREISS R GEWECKE M 1990 Compensation of visually simulated wind drift in the swarming flight of the desert locust (*Schistocerca gregaria*). *J. exp. Biol. 157: 461-482* **flight behav OM**

PRENTØ P 1972 Histochemistry of neurosecretion in the pars intercerebralis-corpus cardiacum system of the desert locust *Schistocerca gregaria. Gen. comp. Endocrinol. 18: 482-500* **NSC brain retroc CC**

PRINGLE JWS 1968 Comparative physiology of the flight motor. *Adv. Insect Physiol. 5: 163-227* **flight aero**

PROUX B PROUX J PHILLIPS JE 1984 Antidiuretic action of corpus cardiacum (CTSH) on long-term fluid absorption across locust recta in vitro. *J. exp. Biol. 113: 409-421* **ADH CC retroc pep**

PROUX J 1976 Variations histologiques du rectum de *Locusta migratoria migratorioides* (Insecte, Orthoptère) après électrocoagulation des cellules neurosécrétrices sous-ocellaires médianes. *Mém. Soc. Sc. phys. nat. Bordeaux 75-76: 29-32* **pep DH NSC brain gut**

PROUX J 1978 Ganglions de la chaîne nerveuse ventrale et équilibre hydrique chez le criquet migrateur. I. Ganglions abdominaux. *Archs. Biol. Bruxelles 89: 297-312* **pep ADH DH NSC abd**

PROUX J 1978 Ganglions de la chaîne nerveuse ventrale et équilibre hydrique chez le criquet migrateur. II. Ganglions sous-oesophagiens. *Archs. Biol. Bruxelles 89: 313-328* **pep ADH DH NSC suboes**

PROUX J 1978 Influence de saignées répétées sur le comportement pondéral et hydrique du criquet migrateur privé de cellules neurosécrétrices sous-ocellaires médianes diurétiques. *C. R. Acad. Sci. Paris Ser. D 286: 981-984* **pep DH NSC brain blood**

PROUX J 1978 Section de la chaîne nerveuse ventrale et comportement hydrique après une saignée chez le criquet migrateur. *C. R. Acad. Sci. Paris Ser. D 287: 305-307* **pep DH NSC blood**

PROUX J 1979 Influence de l'électrocoagulation de la pars intercerebralis et des cellules neurosécrétrices sous-ocellaires médianes sur la comportement pondéral hydrique et alimentaire du criquet migrateur. *Bull. Soc. Zool. Fr. 104: 89-103* **pep DH brain NSC**

PROUX J 1990 La régulation hormonale du métabolisme hydrique chez les criquets grégarisables. *In: Aupelf-Uref J (ed.): La lutte anti-acridienne. Paris: John Libbey Eurotext. 91-101* **rev pep DH ADH**

PROUX J 1990 Recent advances in hormonal regulation of water metabolism in insect. *Prog. Comp. Endocrinol. 342: 541-546* **rev pep DH ADH**

PROUX J BUSCARLET L 1976 Etude, à l'aide d'eau tritiée du rôle des cellules neurosécrétrices sous-ocellairs médianes dans la régulation hydrique des mâles de *Locusta migratoria migratorioides* R. et F. *Acrida 5: 311-320* **pep DH NSC brain**

PROUX J DELAAGE M CHAUVEAU J 1987 Immunoaffinity purification of a diuretic hormone from the nervous ventral cord of the migratory locust. *Comp. Biochem. Physiol. B 88: 807-811* **pep DH**

PROUX J FOURNIER B 1990 The hormonal control of ions and water balance in insects: from the source to the target. *Life Sci. Adv. gen. Endocrinol. 8: 1-15* **rev pep DH ADH**

PROUX J FOURNIER B GIRARDIE J PICQUOT M 1990 Involvement of neuropeptides in insect water regulation: the african locust model. Comparison with other insect and vertebrate systems. *In: Stefano M (ed.), Aspects of neuropeptide function. Manchester: Manchester University Press, 379-395* **pep DH ADH**

PROUX J GIRARDIE A 1982 Neurosecretory regulation of the haemolymphatic titre of a vasopressin-like substance in the migratory locust. *Neurosci. Lett. 33: 73-77* **pep DH**

PROUX J GIRARDIE A 1987 La neurochimie des hormones d'insectes. *Biotechnica 1: 4* **hormX**

PROUX J HERAULT JP 1988 Cyclic AMP: a second messenger of the newly characterised AVP-like insect diuretic hormone, the migratory locust diurectic hormone. *Neuropeptides 12: 7-12* **pep cAMP DH**

PROUX J MILLER C GIRARDIE A 1986 Isolation and purification of two AVP-like factors from *Locusta migratoria. Insect Neurochem. Neurophysiol. 2: 363-366* **pep DH**

PROUX J MILLER C GIRARDIE A DELAAGE M SCHOOLEY D 1986 Isolation and purification of two AVP-like factors from *Locusta migratoria. In: Borkovec AB Gelman DB (eds.), Insect neurochemistry and neurophysiology. Clifton, N.J.: Humana Press. 363-366* **pep DH**

PROUX JP MILLER C LI JP CARNEY RL GIRARDIE A DELAAGE M SCHOOLEY DA 1987 Identification of an arginine vasopressin-like diuretic hormone from *Locusta migratoria.* One is an antiparallel homodimer with diuretic activity. *Biochem. Biophys. Res. Comm. 149: 180-186* **pep DH thor SUBOES cAMP**

PROUX J MILLER C DELAAGE A GIRARDIE A SCHOOLEY D 1986 Purification of the AVP-like locust diuretic hormone. *Bull. Soc. Zool. Fr. 111: 47-48* **pep DH**

PROUX J PICQUOT M HERAULT JP FOURNIER B 1988 Diuretic activity of a newly identified neuropeptide: the AVP-like insect diuretic hormone. Use of a novel bioassay. *J. Insect Physiol. 34: 919-927* **pep DH**

PROUX J PROUX B PHILLIPS J 1985 Source and distribution of factors in locust nervous system which stimulate rectal Cl^- transport. *Can. J. Zool. 63: 37-41* **pep DH**

PROUX J ROUGON-RAPUZZI G 1980 Evidence for vasopressin-like molecule in migratory locust. Radioimmunological measurements in different tissues: correlation with various states of dehydration. *Gen. comp. Endocrinol. 42: 378-383* **pep DH**

PROUX J ROUGON G CUPO A 1982 Enhancement of excretion across locust malpighian tubes by a diuretic vasopressin-like hormone. *Gen. comp. Endocrinol. 47: 449-457* **pep DH**

PROUX J ROUGON G CUPO A GIRARDIE A 1982 Vasopressin-like substance in locust: diuretic function and neuroendocrine regulation of its release. *Gen. comp. Endocrinol. 46: 372-374* **pep DH**

PROUX JP ROUGON-RAPUZZI G REMY C 1980 Influence de la pars intercerebralis et des cellules neurosécrétrices sous-ocellaires médianes sur la teneur en substance apparentée à la vasopressine dans la chaîne nerveuse ventrale et l'hémolymphe du criquet migrateur. Etude radio-immunologique. *J. Physiol. (Paris) 76: 277-282* **pep DH NSC brain**

PUMPHREY RJ 1958 Hearing in man and in animals. *Proc. R. Institute 37: 93-117* **aud cerc abd**

PUMPHREY RJ RAWDON-SMITH AF 1936 Sensitivity of insects to sound. *Nature 137: 900* **aud SO tymp**

QAZI S LUNT GC 1991 Muscarinic receptors in locust ganglia are coupled to phosphatidyl inositol turnover and to adenylate cyclase. *Soc. Neurosci. Abstr. 17: 584* **recept ACh InsP cAMP enz**

QUADRI MAH 1940 On the development of the genitalia and their ducts of Orthopteroid insects. *Trans. R. entomol. Soc. Lond. 90: 121-175* **anat genit devel**

QUEATHAM E 1991 The ontogeny of grasshopper jumping perfomance. *J. Insect Physiol. 37: 129-138* **jump behav devel**

RAABE M 1963 Existence chez divers insectes d'une innervation tritocérébrale des corpora cardiaca. *C. R. Acad. Sci. Paris Ser. D 256: 1552-1555* **retroc brain anat**

RAABE M PANOV AA DAVYDOVA ED CHERVIN D 1979 Neurosecretory products diversity in the pars intercerebralis of insects. *Experientia 35: 404-405* **NSC brain anat**

RADEMAKERS LHPM 1977 Effects of isolation and transplantation of the corpus cardiacum on hormone release from its glandular cells after flight in *Locusta migratoria. Cell Tiss. Res. 184: 213-224* **pep retroc AKH flight CC**

RADEMAKERS LHPM 1977 Identification of a secretomotor centre in the brain of *Locusta migratoria*, controlling the secretory activity of the adipokinetic hormone producing glandular cells of the corpus cardiacum. *Cell Tiss. Res. 184: 381-395* **anat pep retroc AKH brain CC**

RADEMAKERS LHPM BEENAKKERS AMT 1977 Changes in the secretory activity of the glandular lobe of the corpus cardiacum in *Locusta migratoria* induced by flight. *Cell Tiss. Res. 180: 155-171* **pep retroc AKH flight CC**

RAFI F BURTT ET 1974 The role of the neck muscles in moving the head of the locust in different directions in response to striped patterns. *Folia biol. Kracow 22: 289-295* **neck musc OM**

RAFI F BURTT ET 1974 An attempt to find electrophysiologically the visual threshold angle in *Schistocerca gregaria* and its instars in response to a striped pattern. *Folia biol. Kracow 22: 297-302* **DN vis devel**

RAFI F BURTT ET 1975 Measurements of different parameters of the compound eye of the locust, *Schistocerca gregaria*, and related species. *Folia biol. Krakow 23: 115-121* **ret**

RAINEY RC 1960 Applications of theoretical models to the study of flight behaviour in locusts and birds. *In: Beament JWL (ed.), Models and Analogies in Biology, Symp. Soc. Exp. Biol. 14. Cambridge: Cambridge University Press. 122-139* **flight**

RAINEY RC 1963 Metereology and the migration of the desert locust. *World Metereological Organisation: Technical Note No. 54.* **flight**

RAINEY RC 1985 Insect flight: new facts - and old fantasies? *In: Gewecke M Wendler G (eds.), Insect Locomotion. Hamburg & Berlin: Paul Parey Verlag. 241-244* **flight**

RAMING K BREER H 1990 Comparative studies on G-proteins in the nervous tissue of invertebrates. *Comp. Biochem. Physiol. B 95: 861-864* **G-prot**

RAMING K KRIEGER J BREER H 1989 Molecular cloning of an insect pheromone-binding protein. *FEBS Lett. 256: 215-218* **SO chem recept**

RAMING K KRIEGER J BREER H 1990 Molecular cloning, sequencing and expression of cDNA encoding a G_o-protein from insect. *Cellular Signalling 2: 311-321* **G-prot**

RAMIREZ JM 1988 Interneurons in the suboesophageal ganglion of the locust associated with flight initiation. *J. comp. Physiol. A 162: 669-685* **suboes IN flight anat**

RAMIREZ JM FRENCH AS 1990 Phentolamine selectively affects the fast sodium component of sensory adaptation in an insect mechanoreceptor. *J. Neurobiol. 21: 893-899* **oct SO mech mod tox**

RAMIREZ JM ORCHARD I 1990 Octopaminergic modulation of the forewing stretch receptor in the locust *Locusta migratoria. J. exp. Biol. 149: 255-279* **SR DN mod oct flight**

RAMIREZ JM ORCHARD I PEARSON KG 1990 Octopaminergic modulation of sense organs involved in locust flight. *In: Borkovec AB Masler EP (eds.), Insect neurochemistry and neurophysiology. Clifton, N.J.: Humana Press Inc. 421-424* **oct mod SO SR**

RAMIREZ JM PEARSON KG 1988 Generation of motor patterns for walking and flight in motoneurons supplying bifunctional muscles in the locust. *J. Neurobiol. 19: 257-282* **MN IN rhythm flight walk thor anat**

RAMIREZ JM PEARSON KG 1989 Distribution of intersegmental interneurones that can reset the respiratory rhythm of the locust. *J. exp. Biol. 141: 151-176* **IN rhythm resp thor suboes abd anat**

RAMIREZ JM PEARSON KG 1989 Alteration of the respiratory system at the onset of locust flight. I. Abdominal pumping. *J. exp. Biol. 142: 401-424* **flight resp MN IN anat**

RAMIREZ JM PEARSON KG 1990 Octopaminergic modulation of sensory pathways involved in locust flight. *In: Elsner N Roth G (eds.) Brain - Perception - Cognition: Proc. 18th Göttingen Neurobiol. Conf. Stuttgart, New York: Thieme Verlag. Abstract 68* **flight mod oct IN thor**

RAMIREZ JM PEARSON KG 1990 Local anaesthetic action of phentolamine on insect mechanoreceptors. *J. comp. Physiol. A 167: 475-484* **SR mech thor flight oct tox**

RAMIREZ JM PEARSON KG 1990 Chemical deafferentation of the locust flight system by phentolamine. *J. comp. Physiol. A 167: 485-494* **SR thor flight oct tox**

RAMIREZ JM PEARSON KG 1990 Octopamine enhances the excitability of central neurons in the flight system of the locust. *Soc. Neurosci. Abstr. 16: 1091* **flight oct**

RAMIREZ JM PEARSON KG 1991 Octopamine induces bursting and plateau potentials in insect neurones. *Brain Res. 549: 332-337* **oct mod IN thor flight**

RAMIREZ JM PEARSON KG 1991 Octopaminergic modulation of interneurons in the locust flight system. *In: Elsner N Penzlin H (eds.), Synapse-transmission-modulation: Proc. 19th Göttingen Neurobiol. Conf. Stuttgart, New York: Thieme Verlag. Abstract 26* **oct mod IN flight**

RAMIREZ JM PEARSON KG 1991 Octopaminergic modulation of interneurons in the flight system of the locust. *J. Neurophysiol. 66: 1522-1534* **oct mod IN flight thor**

RAMIREZ JM PEARSON KG 1991 Plateau-potentials contribute to the generation of rhythmic depolarizations in locust flight interneurons. *Soc. Neurosci. Abstr. 17: 1580* **oct mod IN flight rhythm thor**

RAMIREZ JM PEARSON KG FRENCH AS 1990 Reversible inactivation of insect mechanoreceptors by phentolamine and its use for chemical deafferentation. *In: Elsner N Roth G (eds.) Brain - Perception - Cognition:*

Proc. 18th Göttingen Neurobiol. Conf. Stuttgart, New York: Thieme Verlag. Abstract 70 **SO inhib mod mech thor tox**

RAMIREZ JM PEARSON KG ORCHARD I 1989 Octopaminergic modulation of neurons involved in the initiation and maintenance of locust flight. *In: Erber J Menzel Pflüger H-J Todt D (eds.), Neural mechanisms of behavior: Proc. 2nd Internat. Congr. Neuroethology. Stuttgart & New York: Thieme Verlag. Abstract 232* **SR DN mod oct flight**

RAMSAY JA 1953 Active transport of potassium by the Malpighian tubules of insects. *J. exp. Biol. 30: 358-369* **blood saline**

RANKIN LL ARBAS EA 1991 Electrophysiological changes that accompany transsynaptically induced atrophy in grasshopper thoracic muscles. *Soc. Neurosci. Abstr. 17: 940* **autot leg musc degen NM**

RAPER JA BASTIANI MJ GOODMAN CS 1983 Pathfinding by neuronal growth cones in grasshopper embryos. I. Divergent choices made by the growth cones of sibling neurons. *J. Neurosci. 3: 20-30* **devel thor IN**

RAPER JA BASTIANI MJ GOODMAN CS 1983 Pathfinding by neuronal growth cones in grasshopper embryos. II. Selective fasciculation onto specific axonal pathways. *J. Neurosci. 3: 31-41* **devel thor IN**

RAPER JA BASTIANI MJ GOODMAN CS 1984 Pathfinding by neuronal growth cones in grasshopper embryos. IV. The effects of ablating the A and P axons upon the behaviour of the G growth cone. *J. Neurosci. 4: 2329-2345* **devel thor IN**

RAPER JA GOODMAN CS 1982 Transient dye coupling between developing neurons reveals patterns of intercellular communication during embryogenesis. *In: Sheffield JB Hilfer RSR (eds.) Cellular communication during ocular development. Berlin, Heidelberg, New York: Springer Verlag. 85-96.* **devel elect**

RAPER JA BASTIANI MJ GOODMAN CS 1984 Guidance of neuronal growth cones: selective fasciculation in the grasshopper embryo. *Cold Spring Harbor Symp. quant. Biol.48: 587-598.* **devel**

RAYNE RC O'SHEA M 1991 Complete processing of a synthetic prohormone of AKH I reconsituted in vivo. *Soc. Neurosci. Abstr. 17: 200* **pep AKH**

RAZIUDDIN M GHOSE IK 1988 Effects of starvation and subsequent feeding on the neuroendocrine system and sexual behaviour of a pyrgomorphid grasshopper, *Poekilocerus pictus* Fabricius. *Ann. Zool. Agra 26: 29-42* **NSC feed pyrg**

RAZIUDDIN M GHOSE IK 1988 Studies on the corpus allatum of the adult female hooded grasshopper, *Teratodes monticollis* (Orthoptera: Acrididae). *J. current Biosci. 5: 24-29* **NSC feed pamph**

RAZIUDDIN M KHAN TR SINGH SB 1977 Neuroendocrine activity in relation to mating and oviposition in the female grasshopper *Poekilocerus pictus. In: Adiyodi KG, Adiyodi Rg (eds.), Advances in invertebrate reproduction, Vol. I; Proc. 1st. Int. Symp. Rep. Physiol. Invertebrates. Kerala, India: Peralam-Kennoth. 319-327.* **NSC behav pyrg ovip**

RAZIUDDIN M KHAN TR SINGH SB 1978 Studies on the neuroendocrine system of the grasshopper *Poekilocerus pictus* Fabr. 1. The protocerebral neurosecretory cells and their axonal pathways in the adult insect. *Folia Morphol. 26: 16-27* **retroc NSC brain anat pyrg**

RAZIUDDIN M KHAN TR SINGH SB 1978 Studies on the neuroendocrine system of the grasshopper *Poekilocerus pictus* Fabr. 2. The corpora cardiaca. *Folia Morphol. 26: 332-337* **retroc CC anat pyrg**

RAZIUDDIN M KHAN TR SINGH SB 1978 Studies on the neuroendocrine system of the grasshopper *Poekilocerus pictus* Fabr. 4. Histomorphology of the corpus allatum in the adult insect. *J. anim. Morphol. Physiol. 26: 61-66* **retroc CAll anat pyrg**

RAZIUDDIN M KHAN TR SINGH SB 1979 Neuropilar neurosecretory fibres in the grasshopper, *Poekilocerus pictus* Fabr. and the cockroach *Periplaneta americana* L. *Zool. Anz. 202: 209-218* **brain anat pyrg dict**

RAZIUDDIN M MEHOTRA PN 1989 Effects of severence of the nerves of the corpora cardiaca and corpora allata on the activity of corpora allata in a pyrgomorphid grasshopper, *Poekilocerus pictus* Fabricius. *Ann. Zool. Agra 27: 1-11* **retroc CAll**

RAZIUDDIN M SINGH SB 1987 Studies on the neuroendocrine system of the grasshopper *Poekilocerus pictus* Fabr. 5. Nervous connections of the corpus allatum. *Zool. Anz. 218: 287-294* **retroc anat CAll pyrg**

RAZIUDDIN M SINGH SB 1989 Studies on the neuroendocrine system during oocyte development in a pyrgomorphid grasshopper, *Poekilocerus pictus* Fabricius. *J. zool. Soc. India 40: 89-102* **retroc NSC pyrg**

REDOUANE K FUZEAU-BRAESCH S 1985 Octopamine sensitive adenylate cyclase in the brain of the migratory locust *Locusta migratoria cinerascens*: comparison between different ages in the solitary and gregarious state. *Comp. Biochem. Physiol. C 81: 451-156* **oct brain enz cAMP phase**

REES D USHERWOOD PNR 1972 Fine structure of normal and degenerating motor axons and nerve-muscle synapses in the locust *Schistocerca gregaria. Comp. Biochem. Physiol. A 43: 83-101* **fineStruc NM MN degen**

REES D USHERWOOD PNR 1972 Effects of denervation on the ultrastructure of insect muscle. *J. Cell Sci. 10: 667-682* **musc fineStruc degen**

REHBEIN H 1973 Experimentell-anatomische Untersuchungen über den Verlauf der Tympanalnervenfasern im Bauchmark von Feldheuschrecken, Laubheuschrecken und Grillen. *Verh. dt. zool. Gesellsch. 1973: 184-189* **aud SO anat thor**

REHBEIN H 1976 Auditory neurons in the ventral cord of the locust: morphological and functional properties. *J. comp. Physiol. A 110: 233-250* **aud IN thor anat**

REHBEIN H KALMRING K RÖMER H 1974 Structure and function of acoustic neurons in the thoracic ventral nerve cord of *Locusta migratoria. J. comp. Physiol. A 95: 263-280* **aud IN SO tymp anat thor**

REHBEIN H KALMRING K SCHWARTZKOPFF J 1974 Verschaltung eines akustischen Neurons im Bauchmark von *Locusta migratoria. Naturwiss. 61: 689* **aud IN anat**

REICHERT H 1985 The cellular basis of sensorimotor coordination in the flight control system of the locust *Locusta migratoria. In: Bush BMH Clarac F (eds.), Coordination of motor behaviour. Soc. Exp. Biol. Seminar Series 24, Cambridge University Press, Cambridge. 121-140* **flight IN MN**

REICHERT H 1988 Constancy and plasticity in the operation of the motor control system for locust flight. *In: Herting G Spatz H-C (eds.), Modulation of synaptic transmission and plasticity in nervous systems. Berlin, Heidelberg, New York: Springer Verlag & NATO scientific affairs Division. 319-322* **pep AKH SO mech SR flight**

REICHERT H 1989 Neural mechanisms underlying axial/appendicular steering reactions in locust flight. *Am. Zool. 29: 161-169* **flight IN MN DN OM rev**

REICHERT H KRENZ W-D 1986 In vivo visualization of individual neurons in arthropod ganglia facilitates intracellular neuropil recording. *J. comp. Physiol. A 158: 625-637* **tech MN IN thor**

REICHERT H MEIER T 1990 Transient expression of cell-specific labels on growth cones and developing synapses in the grasshopper embryo. *Soc. Neurosci. Abstr. 16: 317* **devel DN**

REICHERT H MEIER T FRIEDLI M 1991 Monoclonal antibodies as molecular probes for neuronal recognition molecules. *In: Elsner N Penzlin H (eds.), Synapse-transmission-modulation: Proc. 19th Göttingen Neurobiol. Conf. Stuttgart, New York: Thieme Verlag. Abstract 614* **devel DN**

REICHERT H ROWELL CHF 1985 Integration of non-phaselocked exteroceptive information in the control of rhythmic flight in the locust. *J. Neurophysiol. 53: 1201-1218* **flight IN MN DN OM**

REICHERT H ROWELL CHF 1986 Neuronal circuits controlling flight in the locust: how sensory information is processed for motor control. *Trends in Neurosci. 9: 281-283* **flight IN MN DN OM rev**

REICHERT H ROWELL CHF 1989 Invariance of oscillator interneurone activity during variable motor output by locusts. *J. exp. Biol. 141: 231-239* **flight IN MN DN OM**

REICHERT H ROWELL CHF GRISS C 1985 Course correction circuitry translates feature detection into behavioral action in locusts. *Nature 315: 142-144* **flight IN MN DN OM**

REICHHART JM CHARLET M 1986 Ecdysiotropic activity in brains and corpora cardiaca of larvae and adults of *Locusta migratoria*: in vitro assays. *Int. J. Invert. Reprod. Devel. 10: 17-25* **pep retroc brain CC**

REINECKE M WALTHER C 1978 Aspects of turnover and biogenesis of synaptic vesicles at locust neuromuscular junctions as revealed by zinc iodide-osmium tetroxide (ZIO) reacting with intravesicular SH-groups. *J. Cell Biol. 78: 839-855* **fineStruc NM**

RÉMY C 1984 Immunochemical relationships between vertebrate and invertebrate neuropeptides. *Nova Acta Leopoldina N.F. 56: 99-116* **pep**

RÉMY C DUBOIS MP 1981 Immunohistochemical evidence of methionine enkephalin-like material in the brain of the migratory locust. *Cell Tiss. Res. 218: 271-278* **TS pep enkeph**

RÉMY C GIRARDIE J 1980 Anatomical organisation of two vasopressin-neurophysin-like neurosecretory cells throughout the central nervous system of the migratory locust. *Gen. comp. Endocrinol. 40: 27-35* **NSC pep anat suboes vasop**

RÉMY C GIRARDIE J DUBOIS MP 1979 Vertebrate neuropeptide-like substances in the subesophageal ganglion of two insects: *Locusta migratoria* R. & F. (Orthoptera) and *Bombyx mori* L. (Lepidoptera). Immunocytological investigation. *Gen. comp. Endocrinol. 37: 93-100* **NSC pep suboes anat lep**

RÉMY C GIRARDIE J DUBOIS MP 1979 Des substances apparentées aux neuroblastes de vertébrés révélées par immunofluorescence dans le ganglion sous-oesophagien des insectes. *J. Physiol. (Paris) 75: 1B* **NSC pep suboes anat lep**

RÉMY C VIEILLEMARINGE J 1988 Immunocytology of insect peptides and amines. *In: Thorndyke MC Goldsworthy GJ (eds.): Soc. Exp. Biol. Seminar Series 33. Cambridge: Cambridge University Press. 43-77* **pep tech**

RÉMY C GUY J PELLETIER G BOER HH 1988 Immunohistological demonstration of a substance related to neuropeptide Y and FMRFamide in the cephalic and thoracic nervous systems of the locust *Locusta migratoria. Cell Tiss. Res. 254: 189-195* **NSC pep suboes thor FMRF anat**

REUSE G ZARNACK W 1988 Wingbeat velocities and muscle activity correlated with hindwing movements in locusts. *In: Elsner N Barth FG (eds.) Sense organs: interfaces between environment and behaviour: Proc.16th. Göttingen Neurobiol. Conf. Stuttgart, New York: Thieme Verlag. Abstract 124* **flight musc thor aero**

REUSE G ZARNACK W 1990 Flying locusts adapt to different feedback characteristics. *In: Elsner N Roth G (eds.) Brain - Perception - Cognition: Proc. 18th Göttingen Neurobiol. Conf. Stuttgart, New York: Thieme Verlag. Abstract 73* **flight aero**

REUSE G ZARNACK W 1991 Motor pattern and wing movements controlling the horizontal equilibrium of flying locusts. *In: Elsner N Penzlin H (eds.), Synapse-transmission-modulation: Proc. 19th Göttingen Neurobiol. Conf. Stuttgart, New York: Thieme Verlag. Abstract 86* **flight aero**

REYE DN PEARSON KG 1987 Projections of the wing stretch receptors to central flight neurons in the locust. *J. Neurosci. 7: 2476-2487* **flight SR MN IN anat rhythm propr**

REYE DN PEARSON KG 1988 Entrainment of the locust central flight oscillator by wing stretch receptor stimulation. *J. comp. Physiol. A 162: 77-89* **flight SR MN rhythm propr**

RHEINLAENDER J MÖRCHEN A 1979 "Time-intensity trading" in locust auditory interneurones. *Nature 281: 672-674* **aud IN SO tymp thor**

RICHARDS AG 1963 The ventral diaphragm of insects. *J. Morphol. 113: 17-46* **VD rev**

RIEDE K 1986 Modification of the courtship song by visual stimuli in the grasshopper *Gomphocerus rufus* (Acrididae). *Physiol. Entomol. 11: 61-74* **strid**

RIEDE K 1990 Optically and acoustically evoked escape behaviour in grasshoppers. *In: Elsner N Roth G (eds.) Brain - Perception - Cognition:*

Proc. 18th Göttingen Neurobiol. Conf. Stuttgart, New York: Thieme Verlag. 207 **behav aud vis SO tymp**

RIEDE K KÄMPER G HÖFLER I 1990 Tympana, auditory thresholds, and projection areas of tympanal nerves in singing and silent grasshoppers (Insecta, Acridoidea). *Zoomorphol. 109: 223-230.* **tymp aud anat SO**

RIEGERT PW 1960 The humidity reactions of *Melanoplus bivittatus* (Say) (Orthoptera, Acrididae): antennal sensilla and hygroreception. *Can. Entomol. 92: 561-570* **SO ant hygr behav**

RILEY JR KRUEGER U ADDISON CM GEWECKE M 1988 Visual detection of wind-drift by high-flying insects at night: a laboratory study. *J. comp. Physiol. A 162: 793-798* **flight vis**

RILEY JR REYNOLDS DR 1986 Orientation at night by high-flying insects. *In: Danthanarayana E (ed.), Insect flight. Berlin, Heidelberg, New York: Springer Verlag. 71-87* **flight vis**

RIND FC 1984 A chemical synapse between two motion-detecting neurones in the locust brain. *J. exp. Biol. 110: 143-168* **DN brain opt**

RIND FC 1987 Non-directional, movement sensitive neurons of the locust optic lobe. *J. comp. Physiol. A 161: 477-496* **vis IN opt**

RIND FC 1989 Functional significance of the receptive field organization of the locust DCMDs. Looming detectors? *In: Erber J Menzel R Pflüger H-J Todt D (eds.), Neural mechanisms of behavior: Proc. 2nd Internat. Congr. Neuroethology. Stuttgart & New York: Thieme Verlag. Abstract 177* **DN vis**

RIND FC 1990 A directionally selective motion-detecting neurone in the brain of the locust: physiological and morphological characterization. *J. exp. Biol. 149: 1-19* **flight DN anat OM brain thor**

RIND FC 1990 Identification of directionally selective motion-detecting neurones in the locust lobula and their synaptic connections with an identified descending neurone. *J. exp. Biol. 149: 21-43* **flight DN anat OM brain opt**

RIND FC 1990 Functional significance of the receptive field organization of an identified visual interneurone in the locust. Looming detector? *J. Physiol. (Lond.) 429: 124P* **vis brain**

RIND FC SIMMONS PJ 1990 Neurones in the locust visual system which encode acceleration and are directionally selective for motion in depth. *J. Physiol. (Lond.) 429: 71P* **brain IN vis OM**

RIPLEY SH EWER DW 1953 Neuromuscular facilitation in the locust. *S. Afr. J. Sci. 49: 320-322* **NM**

RITCHIE MG 1990 Are differences in song responsible for assortative mating between subspecies of the grasshopper *Chorthippus parallelus* (Orthoptera: Acrididae)? *Anim. Behav. 39: 685-691* **strid patRec ant olf pher**

ROBB S EVANS PD 1990 FMRFamide-like peptides in the locust: distribution, partial characterization and bioactivity. *J. exp. Biol. 149: 335-359* **pep FMRF**

ROBB S PACKMAN L EVANS PD 1989 Isolation, primary structure, and bioactivity of Schistoflrf, a FMRF-amide-like neuropeptide from the locust, *Schistocerca gregaria. Biochem. Biophys. Res. Comm. 160: 850-856* **pep FMRF**

ROBERT D 1988 Visual steering by flying locusts. Course correction manoeuvres in a flight simulator. *In: Elsner N Barth FG (eds.), Sense organs: interfaces between environment and behaviour: Proc.16th. Göttingen Neurobiol. Conf. Stuttgart, New York: Thieme Verlag. Abstract 120* **flight vis head OM abd**

ROBERT D 1988 Visual steering under closed-loop conditions by flying locusts: flexibility of optomotor response and mechanisms of correctional steering. *J. comp. Physiol. A 164: 15-24* **flight vis OM head abd**

ROBERT D 1989 The ultrasound avoidance behaviour of flying locusts. *In: Elsner N Singer W (eds.), Dynamics and plasticity in neuronal systems: Proc. 17th. Göttingen Neurobiology Conference. Stuttgart, New York: Thieme Verlag. Abstract 145* **flight aud behav HF batev**

ROBERT D 1989 The auditory behaviour of flying locusts. *J. exp. Biol. 147: 279-301* **flight batEv HF aud behav**

ROBERT D 1991 Locusts in a flight simulator: the role of head movements during stable flight and avoidance manoeuvres. *Soc. Neurosci. Abstr. 17: 1244* **flight head behav OM**

ROBERTSON HA 1976 Octopamine, dopamine and noradrenalin content of the brain of the locust *Schistocerca gregaria. Experientia 32: 552-553* **TS brain oct dopa norad**

ROBERTSON RM 1985 Central neuronal interactions in the flight system of the locust. *In: Gewecke M Wendler G (eds.), Insect Locomotion. Hamburg & Berlin: Paul Parey Verlag. 183-194* **IN flight thor**

ROBERTSON RM 1986 Neuronal circuits controlling flight in the locust: central generation of the rhythm. *Trends in Neurosci. 9: 278-280* **flight IN MN rev rhythm thor**

ROBERTSON RM 1988 Insect neurons: synaptic interactions, circuits and the control of behavior. *In: Ali MA (ed.) Nervous systems in invertebrates. New York: Plenum Pess. 393-442* **thor rhythm IN**

ROBERTSON RM 1989 Idiosyncratic computational units generating innate motor patterns: neurones and circuits in the locust flight system. *In: Durbin R Miall RC Mitchinson GM (eds.), The computing neuron. London: Addison-Wesley. 262-277* **IN flight rhythm thor**

ROBERTSON RM 1990 Synchronous activity of flight neurons in the mesothoracic ganglion of the locust. *J. comp. Physiol. A 167: 61-69* **IN flight thor MN**

ROBERTSON RM 1991 Delayed excitatory connections in the flight system of the locust. *J. Neurophysiol. 65: 1150-1157* **flight IN rhythmthor**

ROBERTSON RM 1991 Sensory adaptation: extracellular recording from the locust forewing stretch receptor. *Soc. Neurosci. Abstr. 17: 516* **SO SR tech**

ROBERTSON RM OLBERG RM 1988 A comparison of the activity of flight interneurones in locusts, crickets, dragonflies and mayflies. *Experientia 44: 735-738* **flight IN MN gryll odon ephem thor**

ROBERTSON RM PEARSON KG 1982 A preparation for the intracellular analysis of neuronal activity during flight in the locust. *J. comp. Physiol. A 146: 311-320* **flight tech anat MN thor saline**

ROBERTSON RM PEARSON KG 1983 Interneurons in the flight system of the locust: distribution, connections, and resetting properties. *J. comp. Neurol. 215: 33-50* **flight IN MN anat rhythm thor saline**

ROBERTSON RM PEARSON KG 1984 Interneuronal organization in the flight system of the locust. *J. Insect Physiol. 30: 95-101* **flight IN MN anat rhythm thor**

ROBERTSON RM PEARSON KG 1985 Neural networks controlling locomotion in locusts. *In: Selverston AI (ed.), Model Neural Networks and Behavior. New York: Plenum Press. 21-35* **flight rhythm IN MN rev thor**

ROBERTSON RM PEARSON KG 1985 Neural circuits in the flight system of the locust. *J. Neurophysiol. 53: 110-128* **flight rhythm IN MN anat thor**

ROBERTSON RM PEARSON KG REICHERT H 1982 Flight interneurons in the locust and the origin of insect wings. *Science 217: 177-179* **flight IN MN anat homol thor**

ROBERTSON RM REYE DN 1988 A local circuit interaction in the flight system of the locust. *J. Neurosci. 8: 3929-3936* **IN flight thor nonspik inhib**

ROBERTSON RM WISNIOWSKI L 1988 GABA-like immunoreactivity of identified interneurons in the flight system of the locust, *Locusta migratoria. Cell Tiss. Res. 254: 331-340* **IN thor flight TS GABA**

ROBINSON RN MACALLEN D LUNT GG BATTERSBY MK 1986 The GABA receptor of insect CNS: characterization of a benzodiazapine binding site. *J. Neurochem. 47: 1955-1961* **TS GABA recept**

ROEDER KD 1939 Synchronized activity in the optic and protocerebral ganglia of the grasshopper, *Melanoplus femur-rubrum. J. cell. comp. Physiol. 14:299-307* **opt brain rhythm**

ROEDER KD 1940 The origin of visual rhythms in the grasshopper, *Melanoplus femur-rubrum. J. cell. comp. Physiol. 16: 399-401* **opt rhythm**

ROEDER T 1989 Pharmacology of the octopamine receptor in locust nervous tissue. *In: Elsner N Singer W (eds.), Dynamics and plasticity in neuronal systems: Proc. 17th. Göttingen Neurobiology Conference. Stuttgart, New York: Thieme Verlag. Abstract 58* **TS oct recept**

ROEDER T 1990 High-affinity antagonists of the locust neuronal octopamine receptor. *Eur. J. Pharmacol. 191: 221-224* **TS oct recept**

ROEDER T 1990 Histamine H1-like binding sites in locust nervous system. *In: Elsner N Roth G (eds.) Brain - Perception - Cognition: Proc. 18th Göttingen Neurobiol. Conf. Stuttgart, New York: Thieme Verlag. Abstract 508* **TS hist recept**

ROEDER T 1990 Histamine H1-like binding sites in locust nervous system. *Neurosci. Lett. 116: 331-335* **TS hist recept**

ROEDER T 1991 Pharmacological and molecular characterization of the locust neuronal octopamine receptor. *In: Elsner N Penzlin H (eds.), Synapse-transmission-modulation: Proc. 19th Göttingen Neurobiol. Conf. Stuttgart, New York: Thieme Verlag. Abstract 428* **recept TS oct**

ROEDER T GEWECKE M 1987 Aufnahme von Octopamin durch das Metathorakalganglion und den Extensor-Tibiae-Muskel von *Locusta migratoria. In: Elsner N Creutzfeld O (eds.), New frontiers in brain research: Proc. 15th Göttingen Neurobiol. Conf. Stuttgart, New York: Thieme Verlag. Abstract 251* **TS oct thor leg musc**

ROEDER T GEWECKE M 1988 Octopamine receptors in locusts. *In: Elsner N Barth FG (eds.) Sense organs: interfaces between environment and behaviour: Proc.16th. Göttingen Neurobiol. Conf. Stuttgart, New York: Thieme Verlag p. 370* **TS oct recept**

ROEDER T GEWECKE M 1989 Octopamine uptake systems in thoracic ganglia and leg muscles of *Locusta migratoria. Comp. Biochem. Physiol. C 94: 143-147* **TS oct thor leg musc**

ROEDER T GEWECKE M 1990 Octopamine receptors in locust nervous tissue. *Biochem. Pharmacol. 39: 1793-1797* **TS oct recept**

ROESSINGH P SIMPSON SJ 1984 Volumetric feedback and the control of meal size in *Schistocerca gregaria. Entomol. exp. appl. 36: 279-286* **propr gut feed**

RÖHLER E 1906 Beiträge zur Kenntnis der Sinnesorgane der Insecten. *Zool. Jb. (Anat.) 22: 225-288* **SO ant**

RÖMER H 1976 Die Informationsverarbeitung tympanaler Rezeptorelemente von *Locusta migratoria* (Acrididae, Orthoptera). *J. comp. Physiol. A 109: 101-122* **aud SO tymp**

RÖMER H 1985 Anatomical representation of frequency and intensity in the auditory system of Orthoptera. *In: Kalmring K Elsner N (eds.) Acoustic and vibrational communication in insects. Hamburg & New York: Paul Parey Verlag. 25-32* **SO tymp aud**

RÖMER H BÜNGERS D 1988 Plasticity of the locust auditory pathway following unilateral deafferentation in early larval development. *In: Elsner N Barth FG (eds.), Sense organs: interfaces between environment and behaviour: Proc.16th. Göttingen Neurobiol. Conf. Stuttgart, New York: Thieme Verlag. Abstract 152* **aud devel**

RÖMER H DRONSE R 1982 Synaptic mechanisms of monaural and binaural processing in the locust. *J. Insect Physiol. 28: 365-370* **aud IN**

RÖMER H MARQUART V 1984 Morphology and physiology of auditory interneurons in the metathoracic ganglion of the locust. *J. comp. Physiol. A 155: 249-262* **aud IN thor anat synon**

RÖMER H MARQUART V HARDT M 1988 Organization of a sensory neuropil in the auditory pathway of two groups of Orthoptera. *J. comp. Neurol. 275: 201-215* **SO anat thor aud**

RÖMER H RHEINLAENDER J 1979 Wechsel von De- und Hyperpolarisation in postsynaptischen Strukturen akustischer Interneurone bei *Locusta. Verh. dt. zool. Ges. 1979: 298*

RÖMER H RHEINLAENDER J 1983 Electrical stimulation of the tympanal nerve as a tool for analysing the responses of auditory interneurons in the locust. *J. comp. Physiol. A 152: 289-296* **aud thor IN**

RÖMER H RHEINLAENDER J 1989 Hearing in insects and its adaptation to environmental constraints. *In: Lüttgau H Necker R (eds.), Biological signal processing. Weinheim: VCH Verlag. 146-162* **aud**

RÖMER H RHEINLAENDER J 1989 Hearing in insects and its adaptation to environmental constraints. *In: Lüttgau H Necker R (eds.), Biological signal processing. Weinheim: VCH Verlag. 146-162* **aud**

RÖMER H RHEINLAENDER J DRONSE R 1981 Intracellular studies on auditory processing in the metathoracic ganglion of the locust. *J. comp. Physiol. A 144: 305-312* **aud IN SO thor**

RÖMER H SCHWARTZKOPFF J 1975 Frequenz- und richtungs-spezifische Rezeptoren der Wanderheuschrecke. *Naturwiss. 12: 581* **aud SO tymp**

RÖMER H SEIKOWSKI U 1985 Responses to model songs of auditory neurons in the thoracic ganglia and brain of the locust. *J. comp. Physiol. A 156: 845-860* **aud IN thor suboes brain habit patRec**

ROMEUF M PROUX J RÉMY C 1986 The neurophysin-vasopressin-like system in migratory locust embryos. Immunohistological and radioimmunological studies. *Neurochem. Int. 8: 299-302* **pep neurophys ADH devel brain**

ROMEUF M RÉMY C 1984 Early immunohistochemical detection of somatostatin-like and methionine-enkephalin-like neuropeptides in the brain of the migratory locust embryo. *Cell Tiss. Res. 236: 289-292* **pep enkeph devel brain**

RONACHER B 1987 Feldheuschrecken mit "gesplittetem" Mesothorakalganglion singen und fliegen normal. *In: Elsner N Creutzfeld O (eds.), New frontiers in brain research: Proc. 15th Göttingen Neurobiol. Conf. Stuttgart, New York: Thieme Verlag. Abstract 87* **flight strid rhythm thor**

RONACHER B 1988 Stridulation of acridid grasshoppers: evidence for hemilateral metathoracic oscillators. *In: Elsner N Barth FG (eds.), Sense organs: interfaces between environment and behaviour: Proc.16th. Göttingen Neurobiol. Conf. Stuttgart, New York: Thieme Verlag. Abstract 137* **strid thor rhythm**

RONACHER B 1989 Hemisegmental pattern generators for stridulation of grasshoppers. *In: Erber J Menzel Pflüger H-J Todt D (eds.), Neural mechanisms of behavior: Proc. 2nd Internat. Congr. Neuroethology. Stuttgart & New York: Thieme Verlag. Abstract 67* **strid rhythm thor**

RONACHER B 1989 Stridulation of acridid grasshoppers after hemisection of thoracic ganglia: evidence for hemiganglionic oscillators. *J. comp. Physiol. A 164: 723-736* **strid rhythm thor**

RONACHER B 1990 Neuronal filters as a basis for song pattern recognition in the auditory system of grasshoppers. *In: Elsner N Roth G (eds.) Brain - Perception - Cognition: Proc. 18th Göttingen Neurobiol. Conf. Stuttgart, New York: Thieme Verlag. Abstract 101-110* **aud strid patRec**

RONACHER B 1990 Contributions of brain and thoracic ganglia to the generation of the stridulation pattern in *Chorthippus dorsatus*. *In: Gribakin FG Wiese K Popov AV (eds.), Sensory systems and communication in arthropods. Basel, Boston, Berlin: Birkhäuser Verlag. 317-323* **strid brain thor**

RONACHER B 1991 Contribution of abdominal commissures in the bilateral coordination of the hindlegs during stridulation in the grasshopper *Chorthippus dorsatus. J. comp. Physiol. A 169: 191-201* **strid leg abd thor**

RONACHER B MILLER S 1986 Localization of neuronal pathways involved in two behavioral reactions in a grasshopper. *Naturwiss. 73: 737-738* **strid flight thor**

RONACHER B RÖMER H 1985 Spike synchronization of tympanic receptor fibres in a grasshopper (*Chorthippus biguttulus* L., Acrididae). A possible mechanism for detection of short gaps in model songs. *J. comp. Physiol. A 157: 631-642* **aud SO tymp**

RONACHER B STUMPNER A 1988 Filtering of behaviourally relevant temporal parameters of a grasshopper's song by an auditory interneuron. *J. comp. Physiol. A 163: 517-524* **aud IN thor synon patRec**

RONACHER B VON HELVERSEN D 1991 Stabilizing influence of visual cues on the phonotactic approach during a grasshopper's communication. *In: Elsner N Penzlin H (eds.), Synapse-transmission-modulation: Proc. 19th Göttingen Neurobiol. Conf. Stuttgart, New York: Thieme Verlag. Abstract 138* **vis aud**

RONACHER B VON HELVERSEN D VON HELVERSEN O 1986 Routes and stations in the processing of auditory directional information in the CNS of a grasshopper, as revealed by surgical experiments. *J. comp. Physiol. A 158: 363-374* **aud IN strid thor**

RONACHER B WOLF H 1989 Lesion experiments indicate a hemisegmental organization of the locust flight oscillator. *J. Physiol. (Lond.) 409: 61P.* **flight rhythm thor**

RONACHER B WOLF H REICHERT H 1988 Locust flight behaviour after hemisection of individual thoracic ganglia: evidence for hemiganglionic premotor centers. *J. comp. Physiol. A 163: 749-759* **flight thor rhythm**

ROONWAL ML 1936 Studies of the embryology of the african locust *Locusta migratoria migratorioides* Reich. and Frm. (Orthoptera, Acrididae). I. The early development, with a new theory of multi-phased gastrulation among insects. *Phil. Trans. R. Soc. Lond. (Biol.) 226: 391-421* **devel anat**

ROONWAL ML 1937 Studies of the embryology of the african locust *Locusta migratoria migratorioides* Reich. and Frm. (Orthoptera, Acrididae). II. Organogeny. *Phil. Trans. R. Soc. Lond. (Biol.) 227: 175-224* **devel anat**

ROSCOW JM 1961 The structure, development and variation of the stridulatory apparatus of the species *Stenobothrus lineatus* (Panzer) (Orthoptera: Acrididae). *Proc. R. entomol. Soc. Lond. A 38: 194-199* **anat leg strid brun SO mech**

ROTZLER S 1989 Possible efference to the ocellar system of the locust and its effect on ocellar interneurons. *J. comp. Physiol. A 165: 27-34* **oc IN DN brain**

ROUSSEL J-P 1970 Action de divers facteurs externes sur le rythme cardiaque de *Locusta migratoria* L. *C. R. Acad. Sci. Paris Ser. D 272: 289-292* **heart rhythm**

ROUSSEL J-P 1971 Influence de la pars intercerebralis et des glandes prothoraciques sur le rythme cardiaque de *Locusta migratoria* L. *Archs. Zool. exp. gén. 112: 179-195* **heart rhythm brain**

ROUSSEL J-P 1972 Rythme et régulation du coeur chez *Locusta migratoria migratorioides* L. *Acrida 1: 17-39* **heart rhythm**

ROUSSEL J-P 1972 Innervation du coeur et régulation nerveuse du rythme cardiaque chez *Locusta migratoria* L. *Archs. Zool. exp. gén. 113: 265-294* **heart innerv rhythm**

ROUSSEL J-P 1972 Rôle du ganglion frontal sur le rythme cardiaque de *Locusta migratoria* L. *Experientia 28: 804-805* **SGG heart rhythm**

ROWELL CHF 1961 The structure and function of the prothoracic spine of the desert locust, *Schistocerca gregaria* Forskål. *J. exp. Biol. 38: 457-469* **anat thor IN CO propr leg**

ROWELL CHF 1963 A method for chronically implanting stimulating electrodes into the brain of locusts, and some results of stimulation. *J. exp. Biol. 40: 271-284* **tech brainStim**

ROWELL CHF 1964 Central control of an insect segmental reflex. I. Inhibition by different parts of the nervous system. *J. exp. Biol. 41: 559-572* **groom thor behav**

ROWELL CHF 1965 The control of reflex responsiveness and the integration of behaviour. *In: Treherne JE Beament JWL (eds.), The physiology of the insect central nervous system. New York, London: Academic Press. 157-164* **groom behav rev**

ROWELL CHF 1969 Central control of an insect segmental reflex. II. Analysis of the inhibitory input from the metathoracic ganglion. *J. exp. Biol. 50: 191-202* **groom tech behav thor**

ROWELL CHF 1971 Variable responsiveness of a visual interneurone in the free-moving locust, and its relation to behaviour and arousal. *J. exp. Biol. 55: 727-748* **DN habit behav**

ROWELL CHF 1971 Antennal cleaning, arousal and visual interneurone responsiveness in a locust. *J. exp. Biol. 55: 749-762* **DN habit groom**

ROWELL CHF 1971 The orthopteran descending movement detector (DMD) neurones; a characterisation and review. *Z. vergl. Physiol. 73: 167-194* **vis DN rev**

ROWELL CHF 1974 Boredom and attention in a cell in the locust visual system. *In: Barton Browne L (ed.), Experimental analysis of insect behaviour. Berlin, Heidelberg, New York: Springer Verlag. 87-99* **DN habit vis**

ROWELL CHF 1976 The cells of the insect neurosecretory system: constancy, variability, and the concept of the unique identifiable neuron. *Adv. Insect Physiol. 12: 63-123* **NSC brain suboes thor abd anat rev**

ROWELL CHF 1976 Small system neurophysiology and the study of plasticity. *In: Rozenzweig E Bennett R (eds.), Neural mechanisms of learning and memory. Cambridge, Mass: MIT Press. 463-467* **DN labil rev**

ROWELL CHF 1988 Mechanisms of steering in flight by locusts. *In: Camhi J (ed.), Neuroethology: a multiauthor review. Experientia 44: 389-395* **flight aero IN MN batEv DN**

ROWELL CHF 1989 Descending interneurons of the locust reporting course deviation from flight course: what is their role in steering? *J. exp. Biol. 146: 177-194* **thor IN MN anat DN flight rev**

ROWELL CHF 1989 The taxonomy of invertebrate neurons: a plea for a new field. *Trends in Neurosci. 12: 169-174* **thor IN MN anat**

ROWELL CHF 1991 The organization of exteroceptive sensory inputs to interneurons of the flight neuropil in locusts. *Tissue & Cell 23: 271-276* **thor IN MN anat flight ROWELL CHF DOREY AE** 1967 The number and size of axons in the thoracic connectives of the desert locust, *Schistocerca gregaria* Forsk. *Z. Zellforsch. mikr. Anat. 83: 288-294* **thor anat**

ROWELL CHF HORN G 1967 Response characteristics of neurones in an insect brain. *Nature 216: 702-703* **DN habit**

ROWELL CHF HORN G 1968 Dishabituation and arousal in the response of single nerve cells in an insect brain. *J. exp. Biol. 49: 171-183* **DN habit**

ROWELL CHF McKAY JM 1969 An acridid auditory interneurone. I. Functional connexions and response to single sounds. *J. exp. Biol. 51: 231-246* **aud IN thor**

ROWELL CHF McKAY JM 1969 An acridid auditory interneurone. II. Habituation, variation in response level, and central control. *J. exp. Biol. 51: 247-260* **aud IN thor habit**

ROWELL CHF O'SHEA M 1976 Neuronal basis of a sensory analyser, the acridid movement detector system. I. Effects of simple incremental and decremental stimuli in light- and dark-adapted animals. *J. exp. Biol. 65: 273-288* **DN vis opt ret**

ROWELL CHF O'SHEA M 1976 Neuronal basis of a sensory analyser, the acridid movement detector system. III. Control of response amplitude by tonic lateral inhibition. *J. exp. Biol. 65: 617-625* **DN vis opt**

ROWELL CHF O'SHEA M 1980 Modulation of transmission at an electrical synapse in the locust movement detector system. *J. comp. Physiol. A 137: 233-241* **DN vis opt elect**

ROWELL CHF O'SHEA M WILLIAMS JLD 1977 Neuronal basis of a sensory analyser, the acridid movement detector system. IV. The preference for small field stimuli. *J. exp. Biol. 68: 157-185* **DN vis opt anat**

ROWELL CHF PEARSON KG 1983 Ocellar input to the flight motor system of the locust: structure and function. *J. exp. Biol. 103: 265-288* **flight DN IN MN oc anat**

ROWELL CHF REICHERT H 1985 Compensatory steering in locusts: the integration of non-phaselocked input with a rhythmic motor output. *In: Gewecke M Wendler G (eds.), Insect Locomotion. Hamburg & Berlin: Paul Parey Verlag. 175-182* **flight IN MN OM rev**

ROWELL CHF REICHERT H 1986 Three descending interneurons reporting deviation from course in the locust. II. Physiology. *J. comp. Physiol. A 158: 775-794* **flight DN oc OM wind temp**

ROWELL CHF REICHERT H 1991 Mesothoracic interneurons involved in flight steering in the locust. *Tissue & Cell 23: 75-139* **thor IN MN flight anat synon**

ROWELL CHF REICHERT H BACON J 1985 How locusts fly straight. *In: Barnes WJP Gladden MH (eds.), Feedback and motor control in invertebrates and vertebrates. London, Sydney, Dover, New Hampshire: Croom Helm Ltd. 337-354* **flight DN IN thor OM rev**

ROZHKOVA GI RODIONOVA HI POPOV AV 1984 Two types of information processing in cercal system of insects: directional sensitivity of giant interneurons. *J. comp. Physiol. A 154: 805-815* **cerc SO mech propr inhib abd GF**

RUCK P 1954 Electrical responses of insect dorsal ocelli. *J. cell. comp. Physiol. 44: 527-533* **oc**

RUCK P 1957 The electrical responses of dorsal ocelli in cockroaches and grasshoppers. *J. Insect Physiol. 1: 109-123* **oc dict**

RUNION HI USHERWOOD PNR 1966 New approach to neuromuscular analysis in the intact free-walking insect preparation. *J. Insect Physiol. 12: 1255-1263* **walk**

RUNION HI USHERWOOD PNR 1966 Responsiveness of a mechanoreceptor (chordotonal organ) in the locust leg to static displacement and velocity of stretch. *J. Physiol. (Lond.) 187: 40-41P* **FCO SO propr leg**

RUNION HI USHERWOOD PNR 1968 Tarsal receptors and leg reflexes in the locust and grasshopper. *J. exp. Biol. 49: 421-436* **SO leg propr mech**

RUTHERFORD DM JEFFRIES D LUNT GG WEITZMAN PDJ 1987 GABA binding to receptor sites in locust supraoesophageal ganglia. *Neurosci. Lett. 79: 337-340* **TS GABA recept brain**

RYAN RO ZIEGLER R VAN DER HORST DJ LAW JH 1990 Characterization of apolipophorin III from *Barytettix psolus* and *Melanoplus differentialis*. *Insect Biochem. 20: 127-133* **blood flight metab**

RYCKEBUSCH S LAURENT G 1991 Pilocarpine induces rhythmic activity in leg motor neurons in isolated thoracic ganglia. *Soc. Neurosci. Abstr. 17: 1581* **rhythm tox walk thor ACh TS**

S-RÓZSA K 1974 Relation of the second messenger system to the electrogenesis and regulation of the contraction in the heart cell membrane of Insecta. *Comp. Biochem. Physiol. A 49: 81-88* **heart musc**

S-RÓZSA K KISS T V-SZÖKE I 1973 On the role of bioactive substances in the rhythm regulation of heart muscle cells of Gastropoda and

Insecta. *In: Salánki J (ed..), Invertebrate Neurobiology: Mechanisms of rhythm regulation. Budapest: Kiadó. 167-181* **heart musc**

S-RÓZSA K MILLER TA 1981 Aminergic and peptidergic receptors at the heart muscle of *Locusta migratoria migratorioides* R. & F. *In: S-Rózsa K (ed.), Neurotransmitters in invertebrates. Budapest: Pergamon Press. Adv. Physiol. Sci. 22: 557-581* **pep heart musc TS recept**

S-RÓZSA K V-SZÖKE I 1970 Investigation of the chemical sensitivity of insect hearts. *Annal. Biol. Tihany 37: 99-109* **heart musc**

S-RÓZSA K V-SZÖKE I 1970 Investigation of the chemical sensitivity of insect hearts. *Annal. Biol. Tihany 37: 99-109* **heart musc**

S-RÓZSA K V-SZÖKE I 1971 Characteristics of the resting and action potentials of the heart muscle fibres in some Insecta species. *Annal. Biol. Tihany 38: 65-77* **heart musc**

S-RÓZSA K V-SZÖKE I 1972 Ion mechanisms of the resting and action potentials in the heart of some insect species. *Comp. Biochem. Physiol. A 41: 495-506* **heart musc**

S-RÓZSA K V-SZÖKE I 1975 Pharmacological properties of the receptors at the heart muscle membrane of *Locusta migratoria migratorioides* R. & F. (Insecta). *Annal. Biol. Tihany 42: 99-113* **heart musc**

S-RÓZSA K VERO M 1971 Electrocardiograms in Insecta and Gastropoda and their changes under experimental conditions. *Annal. Biol. Tihany 38: 79-86* **heart musc**

SAAGER F GEWECKE M 1989 Antennal reflexes in the desert locust *Schistocerca gregaria. J. exp. Biol. 147: 519-532* **ant flight MN anat**

SAMARANAYAKA M 1974 The effect of insecticides on osmotic and ionic balance in the locust, *Schistocerca gregaria. Pest. Biochem. Physiol. 7: 510-516* **pep retroc CC**

SAMARANAYAKA M 1976 Possible involvement of monoamines in the release of adipokinetic hormone in the locust, *Schistocerca gregaria. J. exp. Biol. 65: 415-425* **retroc AKH TS pep tox**

SANDEMAN DC 1961 Some aspects of the flight mechanism of the migratory locust, *Locusta migratoria migratorioides* R. & F. *Master's Thesis, Univ. of Natal, Pietermaritzburg, S. Africa.* **flight anat musc**

SANSOM MSP BALL FG KERRY CJ MCGEE R RAMSEY RL USHERWOOD PNR 1989 Markov, fractal, diffusion and related models of ion channel gating - a comparison with experimental data from two ion channels. *Biophys. J. 56: 1229-1243* **TS glu musc channel recept model**

SANSOM MSP USHERWOOD PNR 1990 Single-channel studies of glutamate receptors. *In: Smythies JR Bradley RJ (eds.), Int. Rev. Neurobiol. 32. San Diego, London: Academic Press, Inc. 51-106* **TS glu musc channel recept**

SATELLE DB BREER H 1990 Cholinergic nerve terminals in the central nervous system of insects (Review). *J. Neuroendocrinol. 2: 241* **rev TS ACh**

SATIJA RC SHARMA SP 1967 Histological studies on the deutocerebrum of *Poekilocerus pictus* F. (Orthoptera) and *Mantis religiosa* L. (Dictyoptera). *Res. Bull. Panjab Univ. (n.s.) 17: 233-235* **anat brain ant pyrg dict**

SATIJA RJ 1957 Studies on the course and functions of the giant nerve fibres from the eye and the tegumentary nerve to the ventral nerve cord in *Locusta migratoria. Res. Bull. Panjab Univ. 132: 511-519* **oc windh SO mech vis anat brain IN**

SATIJA RJ 1958 A histological and experimental study of nervous pathways in the brain and thoracic nerve cord of *Locusta migratoria migratorioides. Res. Bull. Panjab Univ. 137: 13-32* **oc windh SO mech vis anat brain IN**

SATTERLIE RA 1985 Structural variability of an identified interneurone in locusts from a wild population. *J. exp. Biol. 114: 691-695* **IN var DN anat flight**

SBRENNA G 1971 Postembryonic growth of the ventral nerve cord in *Schistocerca gregaria* (Orthoptera: Acrididae). *Boll. Zool. 38: 49-74* **anat abd devel**

SCHÄFER M 1991 Using the biocytin method for intracellular labelling of neurons in *Locusta migratoria. In: Elsner N Penzlin H (eds.), Synapse-transmission-modulation: Proc. 19th Göttingen Neurobiol. Conf. Stuttgart, New York: Thieme Verlag. Abstract 352* **DN tech**

SCHÄFER S BICKER G OTTERSEN OP STORM-MATHIESEN J 1988 Taurine-like immunoreactivity in the brain of the honeybee. *J. comp. Neurol. 268: 60-70* **TS taur brain suboes mpt CP anat hym**

SCHIEBE M ORCHARD I WATTS R LANGE AB ATWOOD HL 1990 Characterisation and partial purification of different factors with contraction-potentiating activities from neurohaemal organs of the locust. *J. comp. Neurol. 291: 305-312* **musc mod NSC pep**

SCHIEBE M WALTHER C 1988 Pre- and postsynaptic actions of FMRF-NH2-like peptides in insect neuromuscular synapse. *In: Salánki J S-Rózsa K (eds.), Neurobiology of invertebrates. (Symp. Biol. Hungarica 36). Budapest. 365-376* **NM pep FMRF mod**

SCHIRRMACHER K 1985 Sensorische Eingänge eines identifizierten Bauchmark-Interneurons (A4I1) in der Wanderheuschrecke, *Locusta migratoria. Diplomarbeit, Universität Konstanz, Germany.* **thor windh SO mech IN**

SCHLEMERMEYER E SCHÜTTE M AMMERMÜLLER J 1989 Immunohistochemical and electrophysiological evidence that locust ocellar photoreceptors contain and release histamine. *Neurosci. Lett. 99: 73-78* **oc TS hist ret**

SCHLEMERMEYER E SCHÜTTE M AMMERMÜLLER J 1989 Extraocular photoreceptors in the periocellar epithelium of *Locusta migratoria? Soc. Neurosci. Abstr. 15: 204* **oc ret**

SCHLOTTERER GR 1977 Response of the locust descending movement detector neuron to rapidly approaching and withdrawing visual stimuli. *Can. J. Zool. 55: 1372-1376* **DN habit**

SCHMIDT J RATHMAYER W 1988 Sensory input on common inhibitory neurones in the locust. *In: Elsner N Barth FG (eds.) Sense organs: interfaces between environment and behaviour: Proc.16th. Göttingen Neurobiol. Conf. Stuttgart, New York: Thieme Verlag. Abstract 97* **MN thor nonspik local inhib**

SCHMIDT J RATHMAYER W 1989 Common inhibitory neurones in the locust: input from sense organs of the leg and local interneurones. *In: Elsner N Singer W (eds.), Dynamics and plasticity in neuronal systems: Proc. 17th. Göttingen Neurobiology Conference. Stuttgart, New York: Thieme Verlag. Abstract 133* **MN inhib thor IN nonspik**

SCHMIDT J RATHMAYER W 1989 The inhibited inhibitors: common inhibitory neurons of the locust under sensory and interneuronal control. *In: Erber J Menzel Pflüger H-J Todt D (eds.), Neural mechanisms of behavior: Proc. 2nd Internat. Congr. Neuroethology. Stuttgart & New York: Thieme Verlag. Abstract 15* **MN inhib FCO IN nonspik leg SO propr**

SCHMIDT J ZARNACK W 1987 The motor pattern of locusts during visually induced rolling in long-term flight. *Biol. Cybern. 56: 397-410* **flight MN musc OM**

SCHMITT JB 1954 The nervous system of the pregenital abdominal segments of some Orthoptera. *Ann. entomol. Soc. Am. 47: 677-682* **anat abd**

SCHMITT JB 1959 The cervicothoracic nervous system of a grasshopper. *Smithson. misc. Coll. 137: 307-329* **anat thor neck**

SCHMITT JB 1962 The comparative anatomy of the insect nervous system. *Annu. Rev. Entomol. 7: 137-156* **anat rev**

SCHNEIDER D STEINBRECHT RA 1968 Checklist of insect olfactory sensilla. *Symp. Zool. Soc. Lond. 23: 279-297* **SO ant rev chem fineStruc**

SCHNEIDER H 1990 Distribution of synaptic input of wing stretch receptors in dendrites of locust flight motoneurons. *Soc. Neurosci. Abstr. 16: 1225* **MN flight SR fineStruc**

SCHNEIDER H KUTSCH W 1989 Mapping of distinctive input within an identified motoneurone. *In: Elsner N Singer W (eds.), Dynamics and plasticity in neuronal systems: Proc. 17th. Göttingen Neurobiology Conference. Stuttgart, New York: Thieme Verlag. Abstract 132* **MN flight SR fineStruc**

SCHNEIDER H KUTSCH W 1989 Localization of synaptic contacts and integrative properties of motoneurones: a combined anatomical and physiological study. *In: Erber J Menzel R Pflüger H-J Todt D (eds.), Neural mechanisms of behavior: Proc. 2nd Internat. Congr. Neuroethology. Stuttgart & New York: Thieme Verlag. Abstract 6* **flight MN anat SR**

SCHOLES JH 1964 Discrete subthreshold potentials from the dimly-lit insect eye. *Nature 202: 527-573* **ret**

SCHOOFS L DANGER JM JÉGOU S DE LOOF A VAUDRY H 1990 Localization and characterization of NPY-like immunoreactive peptides in the nervous system and midgut of *Locusta migratoria* and in the brain of *Sarcophaga bullata. Ann. N.Y. Acad. Sci. 611: 513-515.* **pep gut hormX NPY dipt**

SCHOOFS L DANGER JM JÉGOU S PELLETIER G HUYBRECHTS R VAUDRY H DE LOOF A 1988 NPY-like peptides occur in the nervous system and midgut of the migratory locust, *Locusta migratoria*, and in the brain of the grey fleshfly, *Sarcophaga bullata. Peptides (Fayetteville) 9: 1027-1036* **pep gut NPY brain suboes thor abd NSC retroc dipt**

SCHOOFS L HOLMAN GM HAYES TK NACHMAN RJ DE LOOF A 1989 Séparation et identification de peptides du cerveau du criquet migrateur, *Locusta migratoria*, qui influence la contractilité de l'intestin posterieur de la blatte. *Ann. Endocrinol. 50: 49N* **pep gut brain dict**

SCHOOFS L HOLMAN GM HAYES TK NACHMAN RJ DE LOOF A 1990 Locustatachykinin I and II, two novel neuropeptides with homology to peptides of the vertebrate tachykinin family. *FEBS Lett. 261: 397-401* **pep brain retroc suboes loTach**

SCHOOFS L HOLMAN GM HAYES TK NACHMAN RJ DE LOOF A 1990 Isolation, identification and synthesis of locustamyotropin II, an additional neuropeptide of *Locusta migratoria*: member of the cephalomyotropic peptide family. *Insect Biochem. 20: 479-484* **pep brain retroc suboes loMT**

SCHOOFS L HOLMAN GM HAYES TK KOCHANSKY JP NACHMAN RJ DE LOOF A 1990 Locustatachykinin III and IV, two additional insect neuropeptides with homology to peptides of the vertebrate tachykinin family. *Regulatory Peptides 31: 199-212* **pep loTach**

SCHOOFS L HOLMAN GM HAYES TK NACHMAN RJ DE LOOF A 1991 Isolation and identification of Locustapyrokinin, a myotropic peptide from *Locusta migratoria. Gen. comp. Endocrinology 81: 97-104* **pep loPyr**

SCHOOFS L HOLMAN GM HAYES TK NACHMAN RJ DE LOOF A 1990 Isolation and identification of a sulfakinin-like peptide with sequence homology to vertebrate gastrin and cholecystokinin, from the brain of *Locusta migratoria. In: McCaffery A Wilson I (eds.): Chromatography and isolation of insect hormones and pheromones. New York: Plenum Press. 231-241* **pep brain**

SCHOOFS L HOLMAN GM HAYES TK NACHMAN RJ DE LOOF A 1991 Isolation, identification and synthesis of locustomyoinhibiting peptide (LOM-MIP), a novel biologically active neuropeptide from *Locusta migratoria. Regulatory Peptides 36: 111-119* **pep musc LoMIP saline**

SCHOOFS L HOLMAN GM HAYES TK TIPS A NACHMAN RJ VANDESANDE F DELOOF A 1990 Isolation, identification and synthesis of Locustamyotropin (Lom-MT), a novel biologically active insect neuropeptide. *Peptides (Fayetteville) 11: 427-433* **pep loMT**

SCHOOFS L HOLMAN GM PROOST P VAN DAMME J HAYES TK DE LOOF A 1992 Locustakinin, a novel myotropic peptide from the locust, *Locusta migratoria*; isolation, primary structure and synthesis. *Regulatory Peptides 37: 49-57* **loKin pep**

SCHOOFS L ANDERSEN AC EBERLE AN HUYBRECHTS R DE LOOF A VAUDRY H 1988 Coexistence of melanin concentrating hormone and alpha-melanocyte stimulating hormone immunoreactivities in the central nervous system of the locust, *Locusta migratoria. Brain Res. 450: 202-208* **pep brain thor**

SCHOOFS L JÉGOU S TONON MC HUYBRECHTS R VAUDRY H DE LOOF A 1988 Biological, chromatographical and radioimmunological evidence for a melanotropin-like peptide in the central nervous system of *Locusta migratoria. Gen. comp. Endocrinol. 71: 36-44* **pep brain thor**

SCHOOFS L JÉGOU S VAUDRY H VERHAERT P DE LOOF A 1987 Localization of melanotropin-related peptides in the central nervous system of the migratory locust, *Locusta migratoria* and the grey fleshfly, *Sarcophaga bullata. Cell Tiss. Res. 248: 25-31* **pep dipt**

SCHOOFS L JÉGOU S VAUDRY H VERHAERT P DE LOOF A 1987 Proopiomelanocortin and enkephalin related substances in the central nervous system of *Locusta migratoria* and *Sarcophaga bullata. Gen. comp. Endocrinol. 66: 14* **pep dipt**

SCHOOFS L SCHROOTEN S HUYBRECHTS R DE LOOF A 1988 Methionine-enkephalin immunoreactivity in the gonads and nervous system of two insect species: *Locusta migratoria* and *Sarcophaga bullata. Gen. comp. Endocrinol. 69: 1-12* **pep genit dipt**

SCHOOFS L SCHROVEN W DE LOOF A 1990 The occurrence of myotropic factors in the intestine and malpighian tubules of *Locusta migratoria. Comp. Biochem. Physiol. C 97: 195-199* **pep gut musc**

SCHOOLEY DA MILLER CA PROUX JP 1987 Isolation of two arginine vasopressin-like factors from ganglia of *Locusta migratoria. Archs. Insect Biochem. Physiol. 5: 157-166* **pep NSC**

SCHOONEVELD H ROMBERG-PRIVEE HM VEENSTRA JA 1985 Adipokinetic hormone-immunoreactive peptide in the endocrine system of insects: a comparative immunocytochemical approach. *Gen. comp. Endocrinol. 57: 184-194* **pep AKH**

SCHOONEVELD H ROMBERG-PRIVEE HM VEENSTRA JA 1986 Immunocytochemical differentiation between adipokinetic hormone (AKH)-like peptides in neurons and glandular cells in the corpus cardiacum of *Locusta migratoria* and *Periplaneta americana* with C-terminal specific antisera to AKH. *Cell Tiss. Res. 243: 9-14* **pep NSC anat retroc CC AKH dict**

SCHOONEVELD H ROMBERG-PRIVEE HM VEENSTRA JA 1987 Phylogenetic differentiation of glandular cells in corpora cardiaca as studied immunocytochemically with region-specific antisera to AKH. *J. Insect Physiol. 33: 167-176* **pep AKH retroc CC**

SCHOONEVELD H TESSER GI VEENSTRA JA ROMBERG-PRIVEE HM 1983 Adipokinetic hormone and AKH-like peptide demonstrated in the corpora cardiaca and nervous system of *Locusta migratoria* by immunocytochemistry. *Cell Tiss. Res. 230: 67-76* **pep AKH brain retroc CC**

SCHOONEVELD H VEENSTRA JA 1985 Insect neuroendocrine cells and neurons containing adipokinetic hormone (AKH)-immunoreactive substances. *In: Kobayashi H Bern HA Urano A (eds.): Neurosecretion and the biology of neuropeptides. Tokyo: Japan Sci. Soc. Press/ Berlin, Heidelberg, New York: Springer Verlag. 425-434* **pep AKH retroc CC**

SCHOUEST LP MILLER TA OLSEN RW 1988 Quantitative autoradiograhpy of GABA receptors in locust (*Schistocerca americana*) brain. *Pest. Sci. 24: 299-310* **TS GABA brain recept**

SCHULTZ-AELLEN MF ROULET E FISCHER-LOUGHEED J O'SHEA M 1989 Synthesis of a homodimer neurohormone precursor of locust adipokinetic hormone studied by in vitro translation and cDNA cloning. *Neuron 2: 1369-1373* **pep AKH cDNA**

SCHÜRMANN FW 1968 Autoradiographische Untersuchungen zum Ribonucleinsäure- und Protein-Stoffwechsel im Zentralnervensystem von *Locusta migratoria* L. *Z. Zellforsch. mikrosk. Anat. 86: 26-60* **metab**

SCHÜRMANN FW WECHSLER W 1969 Elektronmikroskopische Untersuchung am Antennallobus des Deutocerebrums der Wanderheuschrecke *Locusta migratoria. Z. Zellforsch. mikrosk. Anat. 95: 223-248* **fineStruc ant brain**

SCHÜRMANN FW WECHSLER W 1970 Synapsen im Antennenhügel von *Locusta migratoria. Z. Zellforsch. mikr. Anat. 108: 563-581* **fineStruc ant brain**

SCHWABE J 1906 Beiträge zur Morphologie und Histologie der tympanalen Sinnesapparate der Orthopteren. *Zoologica, Stuttgart 50: 1-154* **anat tymp aud SO gryll tettig**

SCHWENNE T 1990 Kinematische Analyse des Flügelschlags bei Rollmanövern von Wüstenheuschrecken (*Schistocerca gregaria*). *Dissertation, Universität Göttingen, Germany.* **flight aero**

SCHWENNE T ZARNACK W 1987 Movements of the hindwings of *Locusta migratoria*, measured with miniature coils. *J. comp. Physiol. A 160: 657-666* **flight aero**

SCHWENNE T ZARNACK W 1988 Fore- and hindwing movements during roll manoeuvres in locust flight. *In: Elsner N Barth FG (eds.), Sense organs: interfaces between environment and behaviour: Proc.16th. Göttingen Neurobiol. Conf. Stuttgart, New York: Thieme Verlag. Abstract 123* **flight aero OM**

SCHWENNE T ZARNACK W 1989 Kinematical interaction of fore- and hindwing movement during flight manoeuvres in locusts. *In: Elsner N Singer W (eds.), Dynamics and plasticity in neuronal systems: Proc. 17th. Göttingen Neurobiology Conference. Stuttgart, New York: Thieme Verlag. Abstract 117* **flight aero OM**

SCHWENNE T ZARNACK W 1990 Individual wing movements of locusts during roll manoeuvres. *In: Elsner N Roth G (eds.) Brain - Perception - Cognition: Proc. 18th Göttingen Neurobiol. Conf. Stuttgart, New York: Thieme Verlag. Abstract 74* **flight aero OM**

SCHWIENING CJ THOMAS RC 1989 Mechanism of intracellular pH (pH1) regulation by isolated locust neurones. *J. Physiol. (Lond.) 418: 126P* **pH cellcult**

SCHWIENING CJ THOMAS RC 1991 Rapid electroneutral pH regulation in isolated unfed locust neurones can produce large hyperpolarizations. *J. Physiol. (Lond.) 432: 25P* **pH cellcult**

SCHWIENING CJ THOMAS RC 1992 Mechanisms of pHi regulation by locust neurons in isolated ganglia: a microelectrode study. *J. Physiol. (Lond.) 447: 693-709* **pH cellCult**

SCOTT PD HEPBURN HR 1976 Femoral stiffness and jumping in grasshoppers and locusts. *J. Insect Physiol. 22: 913-916* **jump**

SCOTT RH DUCE IR 1987 Pharmacology of GABA receptors on skeletal muscle fibres of the locust (*Schistocerca gregaria*). *Comp. Biochem. Physiol. C 86: 305-311* **GABA TS musc leg recept**

SEABROOK WD 1968 The innervation of the terminal abdominal segments (VIII-XI) of the desert locust *Schistocerca gregaria. Can. Entomol. 100: 693-715* **anat abd genit**

SEABROOK WD 1968 The structure of a pregenital abdominal ganglion of the desert locust *Schistocerca gregaria* (Forskål). *Can. J. Zool. 46: 965-980* **anat abd**

SEABROOK WD 1970 The structure of the terminal ganglionic mass of the locust *Schistocerca gregaria* (Forskål). *J. comp. Neurol. 138: 63-85* **anat abd genit**

SEABROOK WD 1971 An electrophysiological study of the giant fiber system of the locust *Schistocerca gregaria. Can. J. Zool. 49: 555-560* **GF abd IN**

SEATH I 1977 Sensory feedback in the control of mouthpart movement in the desert locust *Schistocerca gregaria. Physiol. Entomol. 2: 147-156* **SO mpt feed propr**

SEATH I 1977 The effects of increasing mandibular load on electrical activity in the mandibular closer muscles in the desert locust *Schistocerca gregaria. Physiol. Entomol. 2: 237-240* **mpt propr SO feed**

SEAVER EC KARLSTROM RO BASTIANI MJ 1991 The restricted spatial and temporal expression of a nervous-system-specific antigen involved in axon outgrowth during development of the grasshopper. *Development 111: 881-893* **devel**

SEMICHON L 1924 L'articulation fémero-tibiale des criquets (Orth.). *Bull. Soc. entomol. Fr. 1924: 162-164* **anat leg**

SHAFIQ SA 1954 Cytological studies of the neurones of *Locusta migratoria.*. Part II. Cytoplasmic inclusions during the differentiation and growth of nerve cells. *Quart. J. micr. Sci. 95: 305-314* **devel**

SHAFIQ SA 1955 Lipochondria in the developing nerve cells of the locust. *J. roy. micr. Soc. 74: 207-208* **devel**

SHANKLAND M 1981 Embryonic development of a sensory afferent projection in the grasshopper embryo. I. Growth of peripheral pioneer axons within the central nervous system. *J. Embryol. exp. Morphol. 64: 169-185* **devel cerc SO abd**

SHANKLAND M 1981 Embryonic development of a sensory afferent projection in the grasshopper embryo. II. Growth and branching of peripheral sensory axons within the central nervous system. *J. Embryol. exp. Morphol. 64: 187-209* **devel cerc SO abd**

SHANKLAND M BENTLEY D 1983 Sensory receptor differentiation and axonal pathfinding in the cercus of the grasshopper embryo. *Dev. Biol. 97: 468-482* **devel cerc abd SO mech**

SHANKLAND M BENTLEY D GOODMAN CS 1982 Afferent innervation shapes the dendritic branching pattern of the medial giant interneuron in grasshopper embryos raised in culture. *Dev. Biol. 92: 507-520* **devel IN GF cerc SO abd cellCult**

SHANKLAND M GOODMAN CS 1982 Development of the dendritic branching pattern of the medial giant interneuron in the grasshopper embryo. *Dev. Biol. 92: 489-506* **devel IN GF cerc SO abd**

SHAW SR 1967 Simultaneous recording from two cells in the locust retina. *Z. vergl. Physiol. 55: 183-194.* **ret**

SHAW SR 1968 Organisation of the locust retina. *Symp. Zool. Soc. Lond. 23: 135-163* **ret rev anat**

SHAW SR 1969 Interreceptor coupling in ommatidia of drone honeybee and locust compound eyes. *Vision Res. 9: 999-1029* **ret hym elect**

SHAW SR 1975 Retinal resistance barriers and electrical lateral inhibition. *Nature 255: 480-483* **ret elect inhib**

SHEPHEARD P 1974 Control of head movement in the locust *Schistocerca gregaria. J. exp. Biol. 60: 735-768* **anat thor suboes MN neck innerv**

SHEPHEARD P 1975 Musculature and innervation of the neck of the desert locust (*Schistocerca gregaria* Forskål). *J. Morphol. 139: 439-464* **anat thor suboes musc neck innerv**

SHEPHERD D BATE CM 1990 Spatial and temporal patterns of neurogenesis in the embryo of the locust (*Schistocerca gregaria*). *Development 108: 83-96* **devel IN thor**

SHEPHERD D LAURENT G 1989 Embryonic development of local spiking neurons in the locust. *Soc. Neurosci. Abstr. 15: 1286* **IN nonspik devel thor local**

SHEPHERD D TYRER NM 1985 Inhibition of GABA uptake potentiates the effects of exogenous GABA on locust skeletal muscle. *Comp. Biochem. Physiol. C 82: 315-321.* **TS GABA musc**

SHULOV A PENER MP 1959 A contribution to knowledge of the development of the egg of *Locusta migratoria migratorioides* (R. & F.). *Locusta (Paris) 6: 73-88* **devel anat**

SIDDIQUI JI KAHN MA 1981 The secretion and perception of a pheromone in the grasshopper *Hieroglyphus nigrorepletus* Bolivar (Orthoptera: Acrididae). *Acrida 10: 233-242* **pher behav ant gut**

SIEGERT KJ MORGAN P MORDUE W 1985 Primary structures of locust adipokinetic hormone II. *Hoppe-Seyler's Z. Physiol. Chem. 366: 723-727.* **pep CC AKH retroc**

SIEGLER MVS 1981 Posture and history of movement determine membrane potential and synaptic events in nonspiking interneurons and motor neurons of the locust. *J. Neurophysiol. 46: 296-309* **local nonspik MN thor**

SIEGLER MVS 1982 Electrical coupling between supernumary motor neurons in the locust. *J. exp. Biol. 101: 105-120* **MN elect**

SIEGLER MVS 1985 Nonspiking interneurons and motor control in insects. *Adv. Insect Physiol. 18: 249-304* **local anat IN thor**

SIEGLER MVS BURROWS M 1979 The morphology of local non-spiking interneurones in the metathoracic ganglion of the locust. *J. comp. Neurol. 183: 121-148* **local nonspik anat thor IN**

SIEGLER MVS BURROWS M 1980 Non-spiking interneurones and local circuits. *Trends in Neurosci. 3: 73-77* **nonspik IN rev local**

SIEGLER MVS BURROWS M 1983 Spiking local interneurones as primary integrators of mechanosensory information in the locust. *J. Neurophysiol. 50: 1281-1295* **local SO mech thor anat**

SIEGLER MVS BURROWS M 1984 The morphology of two groups of spiking local interneurones in the metathoracic ganglion of the locust. *J. comp. Neurol. 224: 463-482* **local anat IN thor**

SIEGLER MVS BURROWS M 1986 Receptive fields of motor neurones underlying local tactile reflexes in the locust. *J. Neurosci. 6: 507-513* **SO local mech IN MN**

SIEGLER MVS MANLEY PE THOMPSON KJ 1991 Sulphide silver staining for endogenous heavy metals reveals subset of dorsal unpaired median DUM neurones in insects. *J. exp. Biol. 157: 565-572* **DUM tech oct**

SIEGLER MVS PHONG MP POUSMAN CA 1991 Motor neurons supplying hindwing muscles of a grasshopper - topography and distribution into anatomical groups. *J. comp. Neurol. 311: 342-355* **MN thor flight anat devel**

SIEGLER MVS POUSMAN CA 1990 Motor neurons of grasshopper metathoracic ganglion occur in stereotypic anatomical groups. *J. comp. Neurol. 297: 297-312* **MN thor leg flight anat devel**

SIEGLER MVS POUSMAN CA 1990 Distribution of motor neurons into anatomical groups in the grasshopper metathoracic ganglion. *J. comp. Neurol. 297: 313-327* **MN thor leg flight anat devel**

SILVER S KALMRING K KÜHNE R 1980 The responses of central acoustic and vibratory interneurons in bushcrickets and locusts to ultrasonic stimulation. *Physiol. Entomol. 5: 427-443* **aud IN vibr HF tettig**

SIMMONS PJ 1980 A locust wind and ocellar brain neurone. *J. exp. Biol. 85: 281-294* **DN oc MN flight**

SIMMONS PJ 1980 Connexions between a movement-detecting visual interneurone and flight motoneurones of a locust. *J. exp. Biol. 86: 87-97* **DN MN flight**

SIMMONS PJ 1981 Synaptic transmission between second- and third-order neurones of a locust ocellus. *J. comp. Physiol. A 145: 265-276* **DN oc brain IN**

SIMMONS PJ 1982 Transmission mediated with and without spikes at connexions between large second-order neurones of locust ocelli. *J. comp. Physiol. A 147: 401-414* **oc DN IN brain**

SIMMONS PJ 1985 Postsynaptic potentials of limited duration in visual neurones of a locust. *J. exp. Biol. 117: 193-213* **oc DN IN brain**

SIMMONS PJ 1986 Interactions made by large second-order neurones of the median ocellus of the locust. *J. comp. Physiol. A 159: 97-105* **oc IN brain**

SIMMONS PJ 1989 Effect of temperature on visual interneurons in an insect. *In: Erber J Menzel Pflüger H-J Todt D (eds.), Neural mechanisms of behavior: Proc. 2nd Internat. Congr. Neuroethology. Stuttgart & New York: Thieme Verlag. Abstract 173* **oc IN brain temp**

SIMMONS PJ 1990 The effects of temperature on locust ocellar L-neurones and their interconnections. *J. comp. Physiol. A 166: 575-583* **oc IN brain temp**

SIMMONS PJ HARDIE RC 1988 Evidence that histamine is a neurotransmitter of photoreceptors in the locust ocellus. *J. exp. Biol. 138: 205-219* **TS oc hist ret**

SIMMONS PJ LITTLEWOOD PMH 1989 Structure of a tonically transmitting synapse between identified interneurones in the locust brain. *J. comp. Neurol. 283: 129-142* **IN brain oc DN fineStruc**

SIMPSON CL CHYB S SIMPSON SJ 1990 Changes in chemoreceptor sensitivity in relation to dietary selection by adult *Locusta migratoria. Entomol. exp. appl. 56: 259-268* **feed SO chem behav**

SIMPSON CL SIMPSON SJ ABISGOLD JD 1990 The role of various amino acids in the protein compensatory response of *Locusta migratoria. In: Proc. 7th Int. Meeting on Insect/Plant Relations. Symp. Biol. Hung. 39: 39-46* **feed**

SIMPSON SJ 1981 An oscillation underlying feeding and a number of other behaviours in fifth instar *Locusta migratoria* nymphs. *Physiol. Entomol. 6: 315-324* **feed rhythm**

SIMPSON SJ 1982 Patterns of feeding: a behavioural analysis using *Locusta migratoria* nymphs. *Physiol. Entomol. 7: 352- 336* **feed**

SIMPSON SJ 1983 The role of volumetric feedback from the hindgut in the regulation of meal size in fifth instar *Locusta migratoria* nymphs. *Physiol. Entomol. 8: 451-467* **feed gut innerv SGG SO propr anat**

SIMPSON SJ 1990 The pattern of feeding. *In: Chapman RF Joern A (eds.), Biology of grasshoppers. New York: J. Wiley & Sons Inc. 73-103* **feed rev**

SIMPSON SJ ABISGOLD JD 1985 Compensation by locusts for change in dietary nutrients: behavioural mechanisms. *Physiol. Entomol. 10: 443-452* **feed**

SIMPSON SJ BERNAYS EA 1983 The regulation of feeding: locusts and blowflies are not SO different from mammals. *Appetite: J. intake Res. 4: 313-346* **feed dipt**

SIMPSON SJ JAMES S SIMMONDS MSJ BLANEY WM 1991 Variation in chemosensitivity and the control of dietary selection behaviour in the locust. *Appetite: J. intake Res. 17: 141-154.* **feed SO chem mpt**

SIMPSON SJ LUDLOW AR 1986 Why locusts start to feed: an analysis of causal factors. *Anim. Behav. 34: 480-496* **feed propr**

SIMPSON SJ SIMMONDS MSJ BLANEY WM 1988 A comparison of dietary selection behaviour in larval *Locusta migratoria* and *Spodoptera littoralis. Physiol. Entomol. 13: 225-238* **feed lep behav**

SIMPSON SJ SIMMONDS MSJ BLANEY WM JONES JP 1989 Compensatory dietary selection occurs in larval *Locusta migratoria* but not *Spodoptera littoralis* after a single deficient meal during ad libitum feeding. *Physiol. Entomol. 15: 235-242* **feed behav lep**

SIMPSON SJ SIMMONDS MSJ WEATHLEY AR BERNAYS EA 1988 The control of meal termination in the locust. *Anim. Behav. 36: 1216-1227* **feed behav**

SIMPSON SJ SIMPSON CL 1989 The mechanisms of compensation by phytophagous insects. *In: Bernays EA (ed.): Insect-Plant Interactions 2. CRC Press, Boca Raton. 111-160* **feed**

SIMPSON SJ WHITE PR 1990 Associative learning and locust feeding: evidence for "learned hunger" for protein. *Anim. Behav. 40: 506-513* **feed behav learn**

SINGH GJP ORCHARD I 1982 Is insecticide-induced release of insect neurohormones a secondary effect of hyperactivity of the central nervous system? *Pest. Biochem. Physiol. 17: 232-242* **TS oct tox**

SINGH GJP ORCHARD I LOUGHTON BG 1981 Octopamine-like action of formamidines on hormone release in the locust *Locusta migratoria. Pest. Biochem. Physiol. 16: 249-255* **TS oct tox**

SINOIR Y 1969 Le rôle des palpes et du labre sans le comportement de prise de nourriture chez la larve du criquet migrateur. *Ann. Nutr. Alim. 23: 167-194* **SO feed mpt**

SINOIR Y 1970 Quelques aspects du comportement de prise de nourriture chez la larve de *Locusta migratoria. Ann. Soc. entomol. Fr. N.S. 6: 391-405.* **SO feed mpt**

SIPPEL M BRECKOW J 1983 Non-linear analysis of the transmission of signals in the auditory system of the migratory locust *Locusta migratoria. Biol. Cybern. 46: 197-205* **tech aud**

SIPPEL M BRECKOW J 1984 Non-monotonic response-intensity characteristics of auditory receptor cells in *Locusta migratoria. J. comp. Physiol. A 155: 633-638* **aud tymp SO**

SKORUPSKI P HUSTERT R 1991 Reflex pathways responsive to depression of the locust coxotrochanteral joint. *J. exp. Biol. 158: 599-605* **leg propr SO**

SKOVMAND O PEDERSEN SB 1978 Tooth impact rate in the song of a short-horned grasshopper: a parameter carrying specific behavioural information. *J. comp. Physiol. A 124: 27-36* **aud patRec strid**

SKOVMAND O PEDERSEN SB 1983 Song recognition and song pattern in a short-horned grasshopper. *J. comp. Physiol. A 153: 393-401* **strid patrec**

SLIFER EH 1935 Morphology and development of the femoral chordotonal organs of *Melanoplus differentialis* (Orthoptera, Acrididae). *J. Morphol. 38: 615-637* **SO propr FCO anat devel leg**

SLIFER EH 1936 The scoloparia of *Melanoplus differentialis* (Orthoptera, Acrididae). *Entomol. News 47: 174-180* **anat SO FCO**

SLIFER EH 1951 Some unusual structures in *Locusta migratoria* and their probable function as thermoreceptors. *Proc. R. Soc. Lond. (Biol.) 138: 414-437* **temp SO head anat**

SLIFER EH 1953 The pattern of specialized heat-sensitive areas on the surface of the body of Acrididae (Orthoptera). I. The males. *Trans. am. entomol. Soc. 79: 37-78* **anat SO temp**

SLIFER EH 1953 The pattern of specialized heat-sensitive areas on the surface of the body of Acrididae (Orthoptera). II.The females. *Trans. am. entomol. Soc. 79: 69-97* **anat SO temp**

SLIFER EH 1954 The variation in size and pattern of the heat-sensitive areas in *Melanoplus keeleri luridus* (Dodge) males (Orthoptera, Acrididae). *Ann. entomol. Soc. Am. 47: 255-264* **anat SO temp**

SLIFER EH 1954 The permeability of the sensory pegs on the antennae of the grasshopper (Orthoptera, Acrididae). *Biol. Bull. 106: 122-128* **SO ant chem tech**

SLIFER EH 1954 The reaction of a grasshopper to an odorous material held near one of its feet. *Proc. R. entomol. Soc Lond. (A) 29: 177-179* **SO leg chem behav rom**

SLIFER EH 1955 The distribution of permeable sensory pegs on the body of the grasshopper (Orthoptera: Acrididae). *Entomol. News 66: 1-5* **SO chem**

SLIFER EH 1955 The detection of odours and water vapour by grasshoppers (Orthoptera, Acrididae) and some new evidence concering the sense organs which may be involved. *J. exp. Zool. 130: 301-317* **SO chem hygr**

SLIFER EH 1956 The response of a grasshopper *Romalea microptera* (Beauvois) to strong odours following amputation of the metathoracic leg at different levels. *Proc. R. entomol. Soc Lond. (A) 31: 95-98* **SO leg chem behav rom**

SLIFER EH 1957 The specialized heat-sensitive areas of the moroccan locust, *Dociostaurus maroccanus* (Thunberg), and of several closely-related species. *Ann. entomol. Soc. Am. 50: 497-499* **anat temp**

SLIFER EH 1960 An abnormal grasshopper with two median ocelli (Orthoptera, Acrididae). *Ann. entomol. Soc. Am. 53: 441-442* **oc anat devel SO**

SLIFER EH 1960 A rapid and sensitive method for identifying permeable areas in the body wall of insects. *Entomol. News 71: 179-182* **SO chem tech**

SLIFER EH 1961 The fine structure of insect sense organs. *Int. Rev. Cytol. 11: 125-159* **SO fineStruc rev**

SLIFER EH 1970 The structure of arthropod chemoreceptors. *Annu. Rev. Entomol. 15: 120-142* **SO rev chem fineStruc**

SLIFER EH 1972 Pores in the thin-walled chemoreceptors of the grasshopper. *Acrida 1: 1-6* **SO chem fineStruc**

SLIFER EH FINLAYSON LH 1956 Muscle receptor organs in grasshoppers and locusts (Orthoptera, Acrididae). *Quart. J. micr. Sci. 97: 617-620* **propr SO musc SR**

SLIFER EH PRESTAGE JJ BEAMS HW 1959 The chemoreceptors and other sense organs on the antennal flagellum of the grasshopper (Orthoptera: Acrididae). *J. Morphol. 105: 145-192* **ant SO fineStruc chem**

SLIFER EH SEKHON SS 1975 The femoral chordotonal organs of a grasshopper, Orthoptera, Acrididae. *J. Neurocytol. 4: 419-438* **anat fineStruc SO propr FCO leg**

SLIFER EH UVAROV BP 1938 Brunner's organ: a structure found on the jumping legs of grasshoppers (Orthoptera). *Proc. R. entomol. Soc Lond. (A) 13: 111-115* **leg SO mech brun**

SMITH TW SIMMONDS MA BLAGBROUGH IS BYCROFT BW MATHER AJ MILLINGTON S USHERWOOD PNR 1990 Monoacylated spermines: antagonists of glutamate receptors. *Eur. J. Pharmacol. 183: 472-473* **TS glu musc tox recept**

SMOLA U 1970 Untersuchung zur Topographie, Mechanik und Strömungsmechanik der Sinneshaare auf dem Kopf der Wanderheuschrecke *Locusta migratoria. Z. vergl. Physiol. 67: 382-402* **windh mech SO**

SMOLA U 1970 Rezeptor- u. Aktionspotentiale der Sinneshaare auf dem Kopf der Wanderheuschrecke *Locusta migratoria. Z. vergl. Physiol. 70: 335-348* **windh mech flight SO**

SNODGRASS RE 1928 Morphology and evolution of the insect head and its appendages. *Smithson. misc. Coll. 81: 158 pp.* **anat musc head**

SNODGRASS RE 1929 The thoracic mechanism of a grasshopper, and its antecedents. *Smithson. misc. Coll. 82: 1-111* **anat musc thor**

SNODGRASS RE 1931 Morphology of the insect abdomen. I. General structure of the abdomen and its appendages. *Smithson. misc. Coll. 85: 1-128* **anat abd**

SNODGRASS RE 1935 The abdominal mechanisms of a grasshopper. *Smithson. misc. Coll. 94: 1-89* **anat musc abd**

SNODGRASS RE 1937 The male genitalia of orthopteroid insects. *Smithson. misc. Coll. 96: 107 pp.* **anat abd genit**

SNODGRASS RE 1960 Facts and theories concerning the insect head. *Smithson. misc. Coll. 142: 1-61* **anat head**

SNOW PM PATEL NH HARRELSON CS GOODMAN CS 1987 Neural-specific carbohydrate moiety shared by many surface glycoproteins in *Drosophila* and grasshopper embryos. *J. Neurosci. 7: 4137-4144* **devel dipt**

SNOW PM ZINN K HARRELSON AL MACALLISTER L SCHILLING J BASTIANI MJ MAKK G GOODMAN CS 1988 Sequence and expression of fasciclin I in grasshopper and *Drosophila*. *Cell 53: 577-587* **devel**

SNOW PM ZINN K HARRELSON AL MCALLISTER L SCHILLING J BASTIANI MJ MAKK G GOODMAN CS 1988 Characterization and cloning of fasciclin I and fasciclin II glycoproteins in the grasshopper. *Proc. nat. Acad. Sci. USA 85: 5291-5295* **devel dipt**

SOBEL EC 1990 The locust's use of motion parallax to measure distance. *J. comp. Physiol. A 167: 579-588* **vis behav**

SOBEL EC 1990 Depth perception by motion parallax and paradoxical parallax in the locust. *Naturwiss. 77: 241-243* **vis behav**

SOKOLIUK T STUMPNER A RONACHER B 1989 GABA-like immunoreactivity suggests an inhibitory function of an auditory interneuron (TN1) in acridid grasshoppers. *In: Elsner N Singer W (eds.), Dynamics* *and plasticity in neuronal systems: Proc. 17th. Göttingen Neurobiology Conference. Stuttgart, New York: Thieme Verlag. Abstract 154* **IN aud GABA TS inhib**

SOKOLIUK T STUMPNER A RONACHER B 1989 GABA-like immunoreactivity suggests an inhibitory function of an auditory interneuron (TN1) in acridid grasshoppers. *Naturwiss. 76: 223-225* **IN aud GABA TS inhib**

SOKOLOV AY 1926 [The structure of the external female genital apparatus in the asiatic locust. The fate of the spermatophores.] (In Russian) *Uchen. Zap. Kazan. Univ. 86: 57-64* **anat abd genit**

SOMBATI S 1983 Orchestrating behaviours: a central modulatory role of octopamine in the locust. *Soc. Neurosci. Abstr. 9: 77* **TS oct mod**

SOMBATI S HOYLE G 1984 Central nervous sensitization and dishabituation of reflex action in an insect by the neuromodulator octopamine. *J. Neurobiol. 15: 455-480* **oct mod**

SOMBATI S HOYLE G 1984 Generation of specific behaviours in a locust by local release into neuropile of natural neuromodulator octopamine. *J. Neurobiol. 15: 481-506* **flight oct TS mod**

SOMBATI S HOYLE G 1984 Glutamatergic central nervous transmission in locusts. *J. Neurobiol. 15: 507-516* **TS glu MN thor**

SOMBATI S LINGLE CS 1985 Properties of single acetylcholine (ACh) receptor channels on dissociated CNS neurons of locust and *Drosophila* . *Biophys. J. Abstr. Vol. 47* **channel recept TS ACh**

SPARKS SL PATTON HD 1968 Evidence for phonoreception in the lubber grasshopper *Romalea microptera. Fedn. Proc. 27: 583* **aud rom**

SPENCER IM CANDY DJ 1974 The effect of flight on the concentrations and composition of haemolymph diacyl glycerols in the desert locust. *Biochem. Soc. Trans. 2: 1093-1096* **metab blood flight lip**

SPENCER IM CANDY DJ 1976 Hormonal control of diacyl glycerol mobilization from fat body of the desert locust *Schistocerca gregaria. Insect Biochem. 6: 289-296* **metab flight hormX lip**

SPORK P PREISS R 1991 Control of flight by means of lateral visual stimuli in the desert locust, *Schistocerca gregaria. In: Elsner N Penzlin H (eds.), Synapse-transmission-modulation: Proc. 19th Göttingen Neurobiol. Conf. Stuttgart, New York: Thieme Verlag. Abstract 90* **flight vis behav**

SPRING JH GÄDE G 1991 Storage and release of neuropeptides from the corpus cardiacum of the eastern lubber grasshopper, *Romalea microptera. J. exp. Zool. 258: 60-68* **rom retroc CC pep**

SPRING JH GÄDE G RINEHART KL 1991 Identification of AKH precursors in *Romalea. Am. Zool. 31: 133A* **pep AKH rom**

SRENG L MOREAU R GIRARDIE A 1990 Locust neuropeptide stimulating sex pheromone production in female european corn borer moth. *J. Insect Physiol. 36: 719-726* **pep pher lep**

STAPLETON A TYRER NM GOOSEY MW COOPER ME 1989 A rapid purification of L-glutamic acid decarboxylase from the brain of the locust *Schistocerca gregaria. J. Neurochem. 53: 1126-1133* **enz glu brain**

STEEVES JD PEARSON KG 1982 Proprioceptive gating of inhibitory pathways to hindleg flexor motoneurons in the locust. *J. comp. Physiol. A 146: 507-515* **jump propr IN thor**

STEEVES JD PEARSON KG 1983 Variability in the structure of an identified interneurone in isogenic clones of locusts. *J. exp. Biol. 103: 47-54* **IN var DN anat**

STEFFENS G BACON JP 1989 Regeneration and synapse reformation of the hindwing stretch receptor with identified motor neurons in *Locusta migratoria. In: Elsner N Singer W (eds.), Dynamics and plasticity in neuronal systems: Proc. 17th. Göttingen Neurobiology Conference. Stuttgart, New York: Thieme Verlag. Abstract 162* **SR regen MN thor**

STEINBRECHT RA 1969 Comparative morphology of olfactory receptors. *In: Pfaffman C (ed.), Olfaction and taste, Proc. 3rd Int. Symp. New York: Rockefeller University Press* . 3-21 **SO chem fineStruc ant**

STEINMANN H 1965 Neuroanatomical studies on the central nervous system of *Acrotylus insubricus* Scop. (Orthoptera). *Budapest: Studia Biologica, Academiae Scientiarum Hungaricae. 142 pp.* **anat**

STELLWAAG F 1916 Wie steuern die Insekten während des Fluges? *Biol. Zentralbl. 36: 30-44* **flight behav aero**

STEPHEN RO BENNET-CLARK HC 1982 The anatomical and mechanical basis of stimulation and frequency analysis in the locust ear. *J. exp. Biol. 99: 279-314* **aud tymp SO**

STEVENSON PA 1989 Central effects of octopamine in the locust. *In: Erber J Menzel Pflüger H-J Todt D (eds.), Neural mechanisms of behavior: Proc. 2nd Internat. Congr. Neuroethology. Stuttgart & New York: Thieme Verlag. Abstract 231* **oct TS rhythm flight mod**

STEVENSON PA KUTSCH W 1986 Basic circuitry of an adult-specific motor program completed with embryogenesis. *Naturwiss. 73: 741-743* **flight TS rhythm devel mod oct**

STEVENSON PA KUTSCH W 1987 A reconsideration of the central pattern generator concept for locust flight. *J. comp. Physiol. A 161: 115-129* **flight TS oct rhythm propr**

STEVENSON PA KUTSCH W 1988 Demonstration of functional connectivity of the flight motor system in all stages of the locust. *J. comp. Physiol. A 162: 247-259* **flight rhythm IN oct thor mod**

STEVENSON PA PFLÜGER H-J ECKERT M RAPUS J 1991 Octopamine immunoreactive neurons in the locust nerve cord. *In: Elsner N Penzlin H (eds.), Synapse-transmission-modulation: Proc. 19th Göttingen Neurobiol. Conf. Stuttgart, New York: Thieme Verlag. Abstract 1* **DUM abd TS oct**

STONE JV MORDUE W 1980 Isolation of insect neuropeptides. *Insect Biochem. 10: 229-239* **pep DH NSC**

STONE JV MORDUE W BATLEY KE MORRIS HR 1976 Structure of locust adipokinetic hormone, a neurohormone that regulates lipid utilisation during flight. *Nature 263: 207-211* **pep flight AKH**

STRANG RHC 1984 Aminoacids as substrates for the locust nervous system. *Proc. 1st Congr. Comp. Physiol. Biochem., Liege. A142-143* **metab**

STRANG RHC CLEMENT EM 1980 The relative importance of glucose and trehalose in the nutrition of the nervous system of the locust *Schistocerca americana gregaria. Insect Biochem. 10: 155-161* **thor metab**

STRANG RHC CLEMENT EM RAE RC 1979 Some aspects of the carbohydrate metabolism of the thoracic ganglia of the locust *Schistocerca gregaria. Comp. Biochem. Physiol. B 62: 217-244* **thor metab**

STRANG RHC WHITTON PS JABBAR A NICHOLSON RA 1989 Taurine as a neuromodulator in the insect. *In: Shain WA (ed.): Functional neurochemistry of taurine. New York: Alan R. Liss Inc. 227-235* **TS taur mod**

STRENGER A 1942 Funktionelle Analyse des Orthopterenkopfes, eine systematisch-funktions-anatomische Studie. *Zool. Jb. (Syst.) 75: 1-72* **anat mpt head musc**

STRICH M-C 1955 Etude de la glande ventrale chez *Locusta migratoria migratorioides* L. (Orth. Acridioidea). *Ann. Sci. natur. (Zool.) 16: 399-411* **anat head**

STRONG L 1965 The relationships between brain, corpora allata, and oocyte growth in the central american locust, *Schistocerca* sp. I. The cerebral neurosecretory system, the corpora allata, and oocyte growth. *J. Insect Physiol. 11: 135-146* **NSC retroc CAll brain anat**

STRONG L 1965 The relationships between brain, corpora allata, and oocyte growth in the central american locust, *Schistocerca* sp. II. The innervation of the corpora allata, the lateral neurosecretory complex, and oocyte growth. *J. Insect Physiol. 11: 271-280* **NSC retroc CAll anat brain**

STROTMANN J TAREILUS E WEGENER J BREER H 1990 Characterisation of olfactory receptor cells in insects: Electrophysiological and immunological approaches. *In: Elsner N Roth G (eds.) Brain - Perception - Cognition: Proc. 18th Göttingen Neurobiol. Conf. Stuttgart, New York: Thieme Verlag. Abstract 310* **ant SO chem**

STUMPNER A 1989 Different aspects of variability in auditory interneurons of a grasshopper (*Chorthippus biguttulus*). *In: Erber J Menzel Pflüger H-J Todt D (eds.), Neural mechanisms of behavior: Proc. 2nd Internat. Congr. Neuroethology. Stuttgart & New York: Thieme Verlag. Abstract 69* **aud IN anat var thor**

STUMPNER A 1989 Physiological variability of auditory neurons in a grasshopper. *Naturwiss. 76: 427-429* **aud IN thor var**

STUMPNER A RONACHER B 1987 Variabilität akustischer Interneurone bei *Chorthippus biguttulus* L.: Probleme für Feldheuschrecke und Experimentator. *In: Elsner N Creutzfeld O (eds.), New frontiers in brain research. Proc. 15th Göttingen Neurobiol. Conf. Stuttgart, New York: Thieme Verlag. Abstract 86* **aud IN thor var**

STUMPNER A RONACHER B 1988 Processing of behaviourally relevant temporal parameters of model songs by *Chorthippus biguttulus*. *In: Elsner N Barth FG (eds.) Sense organs: interfaces between environment and behaviour: Proc.16th. Göttingen Neurobiol. Conf. Stuttgart, New York: Thieme Verlag. Abstract 24* **strid IN thor aud**

STUMPNER A RONACHER B 1991 Auditory interneurons in the metathoracic ganglion of the grasshopper *Chorthippus biguttulus* L.: I. Morphology and physiological characterization. *J. exp. Biol. 158: 391-410* **aud IN thor anat**

STUMPNER A RONACHER B VON HELVERSEN O 1989 Possible contribution of auditory neurons to an IRM of a grasshopper (*Chorthippus biguttulus*). *In: Erber J Menzel Pflüger H-J Todt D (eds.), Neural mechanisms of behavior: Proc. 2nd Internat. Congr. Neuroethology. Stuttgart & New York: Thieme Verlag. Abstract 70* **strid aud IN thor patRec**

STUMPNER A RONACHER B VON HELVERSEN O 1990 Transmission and filtering of behaviourally relevant song parameters by auditory neurons in a grasshopper. *In: Gribakin FG Wiese K Popov AV (eds.), Sensory systems and communication in arthropods. Basel, Boston, Berlin: Birkhäuser Verlag. 351-358* **aud IN thor**

STUMPNER A RONACHER B VON HELVERSEN O 1991 Auditory interneurons in the metathoracic ganglion of the grasshopper *Chorthippus biguttulus* L.: II. Processing of temporal parameters of the male's song. *J. exp. Biol. 158: 411-430* **aud IN thor patRec**

SUGA N 1960 Peripheral mechanism of hearing in locust. *Jap. J. Physiol. 10: 533-546* **aud thor SO tymp**

SUGA N 1963 Central mechanism of hearing and sound localization in insects. *J. Insect Physiol. 9: 867-873* **aud thor IN**

SUGA N KATSUKI Y 1961 Central mechanism of hearing in insects. *J. exp. Biol. 38: 545-558* **aud thor IN**

SUTER C 1986 The action of octopamine and other biogenic amines on locust central neurons. *Comp. Biochem. Physiol. C 84: 181-188* **TS oct**

SUTER C 1986 Does cyclic AMP mediate the action of octopamine on insect neurons ? *Comp. Biochem. Physiol. C 84: 189-194* **TS oct cAMP**

SUTER C USHERWOOD PNR 1985 Action of acetylcholine and antagonists on somata isolated from locust central neurons. *Comp. Biochem. Physiol. C 80: 221-229* **TS ACh cellCult**

SVETLOGORSKAYA ID POPOV AV 1975 [Ultrastructural organisation of the anterior sensory neuropile in the metathoracic ganglion of the locust *Locusta migratoria*]. (in Russian.) *Zh. evol. Biokhim. Fiziol. 11: 419-424* **aud SO fineStruc thor**

SVIDERSKY VL 1965 [The activity of single neurons in the thoracic ganglion in *Locusta migratoria.*] (in Russian.) (in Russian.) *Dok. Akad. Nauk. SSSR 164: 1204-1207* **IN thor**

SVIDERSKY VL 1967 [Electrical activity in receptors concerned with the maintenance of flight in locust.] (in Russian.) *Dok. Akad. Nauk. SSSR 172: 1230-1233* **windh SO mech flight**

SVIDERSKY VL 1967 Afferent mechanisms of the maintenance of flight in locusts. *In: Salanki J (ed.), Symp. Neurobiol. Invert., Tihany 1967. New York: Plenum Press. 479-485* **windh SO mech flight**

SVIDERSKY VL 1967 Central mechanisms controlling the activity of locust flight muscles. *J. Insect Physiol. 13, 899-912* **flight musc**

SVIDERSKY VL 1969 [Nervous control of the fast rhythmic movements of insect muscles (sound producing organs of cicadas and locust flight system).] (in Russian). *Tr. Vses. Entomol. Obsch. SSSR 53: 94-131* **flight rhythm musc**

SVIDERSKY VL 1969 [Neurophysiology of insect flight.] (in Russian.) *Usp. Sovr. Biol. SSSR 67: 465-482* **windh SO mech flight**

SVIDERSKY VL KNJAZEVA NI 1968 [Central processing of impulses sent by the head receptors to the neurons of wing muscles in the locust.] (in Russian.) *Dok. Akad. Nauk. SSSR 183: 486-489* **windh SO mech flight DN**

SWALES LS EVANS PD 1988 Histochemical localization of octopamine- and proctolin-sensitive adenylate cyclase activity in a locust skeletal muscle. *Histochem. 90: 233-239* **TS musc NM oct pep proct rhythm enz cAMP**

SWALES LS EVANS PD 1989 A concentration-dependent localization of octopamine-sensitive adenylate cyclase activity in locust skeletal muscle. *Histochem. 92: 255-260* **TS oct mod cAMP enz musc**

SWINNEN K VANDEN BROECK J VERHAERT P DE LOOF A 1990 Immunocytochemical localization of human growth hormone- and prolactin-like antigenic determinants in the insects, *Locusta migratoria* and *Sarcophaga bullata. Comp. Biochem. Physiol. A 95: 373-378* **pep dipt**

SYSON AJ HUDDART H 1976 The effect of caffeine on excitation-contraction coupling in skeletal and smooth muscle. *J. exp. Biol. 64: 789-798* **musc Ca2**

TAGHERT PH BASTIANI MJ HO RK GOODMAN CS 1982 Guidance of pioneer growth cones: filopodial contacts and coupling revealed with an antibody to Lucifer Yellow. *Dev. Biol. 94: 391-399* **devel tech**

TAGHERT PH DOE CQ GOODMAN CS 1984 Cell determination and regulation during development of neuroblasts and neurones in grasshopper embryo. *Nature 307: 163-165* **devel**

TAGHERT PH GOODMAN CS 1984 Cell determination and differentiation of identified serotonin-immunoreactive neurons in the grasshopper embryo. *J. Neurosci. 4: 989-1000* **5HT abd TS devel**

TAGHERT PH KOTRLA K GOODMAN CS 1982 Cell determination of neuroblasts and neurons during grasshopper embryogenesis. *In: Goodman CS Pearson KG (eds.): Neuronal development: cellular approaches in invertebrates. NRP Bulletin. Boston, Mass.: MIT Press* **devel**

TAKAHASHI R 1925 Postembryonic development of *Pachytylus migratorioides.* [in Japanese] *Zool. Mag. Tokyo 37: 191-203* **devel anat**

TAKEDA S GIRARDIE A 1985 Active principle inducing egg diapause of *Bombyx mori* in corpora cardiaca, corpora allata and suboesophageal ganglion of the locust. *J. Insect Physiol. 31: 761-766* **pep brain retroc suboes diap CC CAll lep**

TAKEDA S GIRARDIE A 1986 An active principle from the corpora cardiaca, corpora allata and suboesophageal ganglion of the locust, inducing egg diapause in *Bombyx mori. Bull. Soc. Zool. Fr. 111: 50-51* **retroc CC CAll suboes diap lep**

TAMARELLE M GIRARDIE J 1986 Permanganate oxidation differentially affects neurophysin and vasopressin immunostaining of locust suboesophageal ganglion and rat neurohypophysis. *Neuroendocr. Lett. 8: 5-8* **suboes neurophys vasop pep**

TAMARELLE M GIRARDIE J 1989 Immunohistochemical investigation of locust neuroparsin-like substances in several insects, in some other invertebrates and vertebrates. *Histochem. 91: 431-435* **pep neuropars**

TAMARELLE M ROMEUF M VANDERHAEGHEN JJ 1988 Immunohistochemical localization of gastrin-colocystokinin-like material in the central nervous system of the migratory locust. *Histochem. 89: 201-207* **pep anat hormX**

TAMARELLE M SALLAMANQUE ML BENEDETTO JP BONNET J VIEILLEMARINGE J GIRARDIE J 1990 Tryptophanyl-tRNA synthetase-like immunoreactivity in the central nervous system and midgut of the migratory locust. Comparisons with gastrin-cholecystokinin-like and octopamine-like immunoreactivity. *Histochem. 95: 195-203* **hormX pep gut enz**

TAREILUS E HANKE W BREER H 1990 Identification of a cationic channel in synaptosomal membranes. *Eur. Biophys. J. 19: 79-86* **channel synaptosome**

TAREILUS E HANKE W BREER H 1990 Neuronal acetylcholine receptor channels from insects: a comparative electrophysiological study. *J. comp. Physiol. A 167: 521-526* **channel ACh TS**

TAYLOR CP 1981 Graded interactions between identified neurons from the simple eyes of an insect. *Brain Res. 215: 382-387* **oc IN brain**

TAYLOR CP 1981 Contribution of compound eyes and ocelli to steering of locusts in flight. I. Behavioral analysis. *J. exp. Biol. 93: 1-18* **flight opt oc vis OM**

TAYLOR CP 1981 Contribution of compound eyes and ocelli to steering of locusts in flight. II. Timing changes in flight motor units. *J. exp. Biol. 93: 19-32* **flight MN opt oc musc propr neck OM**

TAYLOR IR CRESCITELLI F 1944 The electrical changes in response to illumination of the dark- and light-adapted eye of *Dissosteira carolina. Physiol. Zool. 17: 193-199* **ret**

THEOPHILIDIS G 1982 A simple method for studying the innervation of a complex muscle. *J. exp. Biol. 96: 435-438* **tech innerv musc leg**

THEOPHILIDIS G 1983 A comparative study of the anatomy and innervation of the metathoracic extensor tibiae muscle in three orthopteran species. *Comp. Biochem. Physiol. A 75: 285-292* **anat innerv leg musc jump**

THEOPHILIDIS G 1988 The study of action potential propagation failure at the branching point of an insect motor axon. *Comp. Biochem. Physiol. A 90: 257-263* **MN**

THEOPHILIDIS G BURNS M 1983 The innervation of the mesothoracic flexor tibiae muscle of the locust. *J. exp. Biol. 105: 373-388* **anat innerv MN inhib DUM leg musc**

THEOPHILIDIS G BURNS MD 1979 A muscle tension receptor in the locust leg. *J. comp. Physiol. A 131: 247-254* **propr leg anat SO**

THEOPHILIDIS G BURNS MD 1982 A gold-plated suction electrode for extracellular recording and dye infusion. *J. exp. Biol. 98: 455-457* **tech**

THEOPHILIDIS G BURNS MD 1990 The firing pattern of the locust (*Schistocerca gregaria americana*) mesothoracic femoral motor axons in resistance reflexes and during walking on a treadmill. *J. Insect Physiol. 36: 513-522* **MN walk reflex thor**

THEOPHILIDIS G DIMITRIADIS V 1990 The structure and innervation of the metathoracic flexor tibiae muscle of two species of Orthoptera (Insecta). *Comp. Biochem. Physiol. A 97: 583-594* **anat innerv leg musc jump**

THEUNIS W BLÄHSER S HUYBRECHTRS R DE LOOF A 1989 Luteinising hormone, follicle stimulating hormone and gonadotropin releasing hormone immunoreactivity in two insects: *Locusta migratoria* R. & F. and *Sarcophaga bullata* (Parker). *Int. J. Invert. Reprod. Devel. 16 : 111-117* **pep anat dipt**

THEUNIS W HUYBRECHTS R DE LOOF A 1990 Partial purification from locust cephalic neuronal tissue and ventral nerve cord of factors influencing the contractility of the isolated oviduct of *Locusta migratoria*. *Comp. Biochem. Physiol. C 95: 31-34* **hormX brain musc genit**

THOMAS JB BASTIANI MJ BATE CM GOODMAN CS 1984 From grasshopper to *Drosophila* : a common plan for neuronal development. *Nature 310: 203-207* **anat var devel**

THOMAS JG 1952 A comparison of the pterothoracic skeleton and flight muscles of male and female *Lamarkiana* sp. (Orthoptera, Acrididae). *Proc. R. entomol. Soc. Lond. (A) 27: 1-12* **anat musc thor flight pamph**

THOMAS JG 1953 A comparison of the flight muscles of Acrididae with different wing development. *Proc. R. entomol. Soc. Lond. (A) 28: 47-56* **anat musc flight thor**

THOMAS JG 1954 The post-embryonic development of the dorsal part of the pterothoracic skeleton and certain muscles of *Locusta migratoria migratorioides* (Reiche & Fairm.). *Proc. zool. Soc. Lond. 124: 229-238* **anat musc flight devel thor**

THOMAS JG 1963 *Dissection of the locust. London: H.F. & G. Witherby Ltd. 72 pp.* **anat**

THOMAS JG 1965 The abdomen of the female desert locust (*Schistocerca gregaria* Forskål) with special reference to the sense organs. *Anti-Locust Bull. 42. London: HMSO. 20 pp.* **anat SO abd genit**

THOMAS JG 1966 Sense organs on the mouthparts of the desert locust (*Schistocerca gregaria*). *J. Zool. (Lond.) 148: 420-448* **anat mpt SO**

THOMPSON KJ 1986 Oviposition digging in the grasshopper. I. Functional anatomy and the motor programme. *J. exp. Biol. 122: 387-411* **anat genit rhythm ovip**

THOMPSON KJ 1986 Oviposition digging in the grasshopper. II. Descending neural control. *J. exp. Biol. 122: 413-425* **genit DN ovip**

THOMPSON KJ SCHABTACH E 1983 Post-embryonic development of grasshopper ovipositors and the oviposition motor program. *Soc. Neurosci. Abstr. 9: 242* **genit devel rhythm ovip**

THOMPSON KJ SIEGLER MVS 1989 Properties of the small dorsal unpaired median (DUM) neurons of the grasshopper. *Soc. Neurosci. Abstr. 15: 1296* **DUM thor**

THOMPSON KJ SIEGLER MVS 1991 Anatomy and physiology of spiking local and intersegmental interneurons in the median neuroblast lineage of the grasshopper. *J. comp. Neurol. 305: 659-675* **IN thor DUM GABA TS aud wind devel local**

THOMPSON KJ SIEGLER MVS 1991 Three classes of neuron in the median neuroblast lineage of the grasshopper. *Soc. Neurosci. Abstr. 17: 1478* **thor IN devel DUM oct TS aud flight local**

THOMPSON KSJ BACON JP 1991 The vasopressin-like immunoreactive (VPLI) neurons of the locust, *Locusta migratoria*: II. Physiology. *J. comp. Physiol. A 168: 619-630* **suboes NSC pep vasop FMRF anat**

THOMPSON KSJ TYRER NM MAY ST BACON JP 1991 The vasopressin-like immunoreactive (VPLI) neurons of the locust, *Locusta migratoria*: I. Anatomy. *J. comp. Physiol. A 168: 605-617* **suboes NSC pep vasop FMRF anat**

THORSON J 1964 Dynamics of motion perception in the desert locust. *Science 145: 69-71* **vis head neck OM**

THORSON J 1966 Small-signal analysis of a visual reflex in the locust. I. Input parameters. *Kybernetik 3: 41-53* **vis head neck OM**

THORSON J 1966 Small-signal analysis of a visual reflex in the locust. II. Frequency dependence. *Kybernetik 3: 54-66* **vis head neck OM**

THÜRING DA 1986 Variability of motor output during flight steering in locusts. *J. comp. Physiol. A 156: 655-663* **flight MN OM musc**

TIEGS OW 1955 The flight muscles of insects - their anatomy and histology; with some observations on the structure of striated muscle in general. *Phil. Trans. R. Soc. Lond. (Biol.) 238: 221-348* **musc fineStruc anat flight**

TIETZ HM 1923 The anatomy of the digestive system of the Carolina locust (*Dissosteira carolina* L.). *Ann. entomol. Soc. Am. 16: 256-268* **anat gut**

TITMUS MJ HOYLE G 1977 Morphology of identified neuromuscular junctions in locust jumping muscle. *Soc. Neurosci. Abstr. 7: 595* **NM anat fineStruc**

TOBE SS GIRARDIE J GIRARDIE A 1982 Enhancement of juvenile hormone biosysnthesis in locusts following electrostimulation of cerebral neurosecretory cells. *J. Insect Physiol. 28: 867-871* **NSC brain retroc brainStim CAll JH**

TOBIAS JM 1948 The high potassium and low sodium in the body fluid and tissues of a phytophagous insect, the silkworm *Bombyx mori* and the changes before pupation. *J. cell. comp. Physiol. 52: 131-147* **blood rom lep**

TORTAJADA JM 1960 Sobre la estructura del cerebro en los acridios. *Acridium aegyptium y Calliptamus italims.* *Trab. Inst. Cajal Invest. biol. 52: 131-147* **anat brain opt MB CB**

TSANG KR FREEEMAN FA KURTTI TJ BROOKS MA 1981 New cell lines from embryos of *Melanoplus sanguinipes* (F.) (Orthoptera: Acrididae). *Acrida 10: 105-112* **cellCult tech devel**

TSUKAHARA Y HORRIDGE GA 1977 Miniature potentials, light adaptation and after potentials in locust retinula cells. *J. exp. Biol. 68: 137-149* **ret**

TSVILENEVA VA 1953 [On the structure of the ventral nerve cord of the asiatic locust] (in Russian). *Izv. Otdel. estest. Nauk., Akad. Nauk. Tadzhik SSR 2: 49-55* **anat thor abd**

TÜMPEL R 1914 Bau und Wirkungsweise der Punktaugen bei *Acridium aegypticum* L. *Z. wiss. Insektbiol. 10: 275-282* **oc anat**

TUNSTALL J HORRIDGE GA 1967 Electrophysiological investigation of the optics of the locust retina. *Z. vergl. Physiol. 55: 167-182* **ret**

TYRER NM 1969 Time course of contraction and relaxation in embryonic locust muscle. *Nature 224: 815-817* **musc devel**

TYRER NM 1970 Quantitative estimation of the stage of development of the locust embryo. *J. Embryol. exp. Morphol. 23: 705-718* **devel anat**

TYRER NM 1971 Innervation of the abdominal intersegmental muscles in the grasshopper. I. Axon counts using unconventional techniques for the electron microscope. *J. exp. Biol. 55: 305-314* **anat abd musc flight resp MN innerv**

TYRER NM 1971 Innervation of the abdominal intersegment muscles in the grasshopper. II. Physiological analysis. *J. exp. Biol. 55: 315-324* **abd musc innerv**

TYRER NM 1973 Functional development of the reticular system in an insect muscle with synchronously differentiating cells. *J. Cell Sci. 12: 197-215* **musc devel**

TYRER NM 1981 Transmission of wind information on the head of the locust to flight motor neurons. *In: Salánki J (ed.), Neurobiology of the Invertebrates; Adv. Physiol. Sci. 23. Oxford: Pergamon Press and Budapest: Akadémiai Kiadó. 557-571* **DN flight MN**

TYRER NM 1983 Locust flight neurones within the anatomical framework of the thoracic ganglia. *In: Nachtigall W (ed.), BIONA-report 2 "Insekten Flug"; Akad. Wiss. Mainz. Stuttgart & New York: Gustav Fischer Verlag. 105-112* **flight MN anat SR**

TYRER NM ALTMAN JS 1974 Motor and sensory flight neurones in a locust demonstrated using cobalt chloride. *J. comp. Neurol. 157: 117-138* **anat MN SO SR flight propr thor**

TYRER NM ALTMAN JS 1976 Spatial distribution of synapses onto thoracic motor neurones in locusts. *Cell Tiss. Res. 166: 389-398* **flight MN fineStruc**

TYRER NM ALTMAN JS 1980 Identification of cobalt-filled neurones in sections (light and electron microscopy). *In: Strausfeld NJ Miller TA (eds.), Neuroanatomical techniques. Insect nervous system. Berlin, Heidelberg, New York: Springer Verlag. 429-446* **tech anat**

TYRER NM BACON JP DAVIES CA 1976 Do wind-sensitive head hairs in the locust have a spatial representation in the CNS? *In: Broughton WB (ed.), Insect neuroacoustics. London: City of London Polytechnic Press. 37-45* **anat SO windh mech**

TYRER NM BACON JP DAVIES CA 1979 Sensory projections from the wind-sensitive hairs of the locust *Schistocerca gregaria*. *Cell Tiss. Res. 203: 79-92* **anat SO windh mech**

TYRER NM BELL EM 1974 The intensification of cobalt-filled neurone profiles using a modification of Timm's sulphide-silver method. *Brain Res. 73: 151-154* **tech**

TYRER NM GREGORY GE 1982 A guide to the anatomy of locust suboesophageal and thoracic ganglia. *Phil. Trans. R. Soc. Lond. (Biol.) 297: 91-123* **anat suboes thor**

TYRER NM POZZA MF HUMBEL U BACON JP 1988 The "tritocerebral commissure dwarf" (TCD): a major GABA-immunoreactive descending interneuron in the locust. *J. comp. Physiol. A 164: 141-150* **flight DN windh SO mech ant TS GABA anat**

TYRER NM TURNER JD ALTMAN JS 1984 Identifiable neurons in the locust central nervous system that react with antibodies to serotonin. *J. comp. Neurol. 227: 313-330* **anat 5HT brain suboes thor abd CO mpt saliv opt TS**

UCHIYAMA H KATSUKI Y 1956 Recording of action potentials from the antennal nerve of locusts by means of microelectrodes. *Physiol. comp. œcol. 4: 154-163* **ant SO saline tech**

USHERWOOD PNR 1962 The action of the alkaloid ryanodine on insect skeletal muscle. *Comp. Biochem. Physiol. 6: 181-199* **channel Ca2 musc**

USHERWOOD PNR 1963 Response of insect muscle to denervation. I. Resting potential changes. *J. Insect Physiol. 9: 247-255* **musc degen**

USHERWOOD PNR 1963 Response of insect muscle to denervation. II. Changes in neuromuscular transmission. *J. Insect Physiol. 9: 811-825* **musc degen NM**

USHERWOOD PNR 1963 Spontaneous miniature potentials from insect muscle fibres. *J. Physiol. (Lond.) 169: 149-160* **NM**

USHERWOOD PNR 1967 Insect neuromuscular mechanisms. *Am. Zool. 7: 553-582* **TS musc NM rev**

USHERWOOD PNR 1967 Permeability of insect muscle fibres to potassium and chloride ions. *J. Physiol. (Lond.) 191: 29-30P* **musc channel**

USHERWOOD PNR 1968 A critical study of the evidence for peripheral inhibitory axons in insects. *J. exp. Biol. 49: 201-222* **musc NM inhib innerv**

USHERWOOD PNR 1969 Glutamate sensitivity of denervated insect muscle fibres. *Nature 223: 411-413* **TS glu NM degen**

USHERWOOD PNR 1972 Transmitter release from insect excitatory motor nerve terminals. *J. Physiol. (Lond.) 227: 527-552* **TS Glu leg NM MN**

USHERWOOD PNR 1973 Action of iontophoretically applied gamma-aminobutyric acid on locust muscle fibres. *Comp. Biochem. Physiol. A 44: 663-664* **TS GABA musc**

USHERWOOD PNR 1973 Release of transmitter from degenerating locust motoneurones. *J. exp. Biol. 59: 1-16* **musc TS degen glu**

USHERWOOD PNR 1973 Glutamate as an excitatory neuromuscular transmitter in insects. *Neurosci. Res. Prog. Bull. 10: 136-143* **TS glu NM**

USHERWOOD PNR 1974 Nerve-muscle transmission. *In: Treherne JE (ed.): Insect neurobiology. Amsterdam, Oxford: North-Holland. 245-305* **NM**

USHERWOOD PNR 1976 Aminoacid receptors on insect muscle. *In: Reuben JP Purpura DP Bennett MVL Kandel ER (eds.): Electrobiology of nerve, synapse and muscle. New York: Raven Press. 81-92* **TS recept musc glu recept**

USHERWOOD PNR 1976 Transmitter release from insect motoneurons. *In: Spencer Davies P (ed.): Perspectives in experimental biology. Vol.1: Zoology. Oxford & New York: Pergamon Press. 353-360* **TS glu MN NM**

USHERWOOD PNR 1979 Glutamate receptors and glutamate-gated ionophores of locust muscle. *In: Kuczera JH Gabrielska J Przestalski S (eds.): 5th Winter School on Biophysics of membrane transport, Vol. III. Wroclaw, Poland: Agricultural University Press. 77-102* **TS glu NM recept channel**

USHERWOOD PNR 1980 Neuromuscular transmitter receptors of insect muscle. *In: Sattelle DB Hall LM Hildebrand JG (eds.): Receptors for neurotransmitters, hormones and pheromones in insects. Amsterdam: Elsevier/North-Holland Biomedical Press. 141-152* **TS glu NM recept**

USHERWOOD PNR 1980 Peripheral glutamatergic synapses in insects. *In: Taxi J (ed.), Ontogenesis and functional mechanisms of peripheral synapses. Amsterdam: Elsevier/North-Holland Biomedical Press. 367-383* **TS glu NM**

USHERWOOD PNR 1981 Glutamate synapses and receptors on insect muscle. *In: Di Chiara G Gessa GL (eds.): Glutamate as a neurotransmitter. New York: Raven Press. 183-193* **TS glu musc recept NM**

USHERWOOD PNR 1981 Amino-acid gated ion channels in locust muscle membrane. *In: Ròzsa KS (ed.): Neurotransmitters in invertebrates. Oxford & New York: Pergamon Press. Adv. Physiol. Sci. 22: 191-203* **TS glu channel musc**

USHERWOOD PNR CLARK RB GRATION KAF PATLAK JB 1979 Glutamatergic synapses and glutamate receptors on locust muscle. *J. Physiol. (Paris) 75: 615-621* **glu musc recept NM**

USHERWOOD PNR COCHRANE DG REES D 1968 Changes in structural, physiological and pharmalogical properties of insect excitatory nerve-muscle synapses after motor nerve section. *Nature 218: 589-591* **TS glu NM degen**

USHERWOOD PNR CULL-CANDY SG 1974 Distribution of glutamate sensitivity on insect muscle fibres. *Neuropharmacol. 13: 455-461* **NM TS glu leg**

USHERWOOD PNR CULL-CANDY SG 1975 Pharmacology of somatic muscle. *In: Usherwood PNR (ed.), Insect muscle. New York, London: Academic Press. 207-280* **NM TS glu leg**

USHERWOOD PNR DUCE IR BODEN P 1983 Slowly-reversible block of glutamate receptor-channels by venoms of the spiders, *Argiope trifasciata* and *Araneus gemma. J. Physiol. (Paris) 79: 241-245* **TS glu musc channel recept**

USHERWOOD PNR DUCE IR DONALDSON PL 1982 A possible mechanism for exitant aminoacid toxicity. *In: Yoshida H Hagihara Y Ebashi S (eds.): Advances in pharmacology and therapeutics II. Oxford & New York: Pergamon Press. 277-281* **TS glu tox**

USHERWOOD PNR GRUNDFEST H 1964 Inhibitory post-synaptic potentials in grasshopper muscle. *Science 143: 817-818* **TS GABA NM inhib leg musc rom**

USHERWOOD PNR GRUNDFEST H 1965 Peripheral inhibition in skeletal muscle of insects. *J. Neurophysiol. 28: 497-518* **musc leg TS inhib GABA NM rom**

USHERWOOD PNR MACHILI P 1966 Chemical transmission at the insect excitatory neuromuscular synapse. *Nature 210: 634-636* **TS glu NM**

USHERWOOD PNR MACHILI P 1968 Pharmacological properties of excitatory neuromuscular synapses in the locust. *J. exp. Biol. 49: 341-361* **NM TS glu leg**

USHERWOOD PNR MACHILI P LEAF G 1968 L-Glutamate at insect excitatory nerve-muscle synapses. *Nature 219: 1169-1172* **NM TS glu leg**

USHERWOOD PNR REES D 1972 Quantitative studies of the spatial distribution of synaptic vesicles within normal and degenerating motor axons of the locust, *Schistocerca gregaria. Comp. Biochem. Physiol. A 43: 103-118* **musc fineStruc degen**

USHERWOOD PNR RUNION HI 1970 Analysis of the mechanical response of metathoracic extensor tibiae muscles of free-walking locusts. *J. exp. Biol. 52: 39-58* **walk subg jump musc leg**

USHERWOOD PNR RUNION HI CAMPELL JI 1968 Structure and physiology of a chordotonal organ in the locust leg. *J. exp. Biol. 48: 305-323* **FCO leg SO propr anat**

UVAROV B 1966 *Grasshoppers and locusts. Vol. 1, Anatomy, physiology, development, phase polymorphism, introduction to taxonomy. London: Cambridge University Press. 481 pp.* **anat phase rev**

UVAROV B 1977 *Grasshoppers and locusts. Vol. 2, Behaviour, ecology, biogeography, population dynamics. London, Centre for Overseas Pest Research. 613 pp.* **behav rev**

VAANDRAGER SH HALLER TB VAN MARREWIJK WJA BEENAKKERS AMT 1989 Fractionation and kinetic properties of trehalase from flight muscles and haemolymph of the locust, *Locusta migratoria. Insect Biochem. 19: 89-94* **enz musc flight blood**

VAANDRAGER SH VAN MARREWIJK WJA BEENAKKERS AMT 1986 Glycogen phosphorylase activity in flight muscles of *Locusta migratoria* at rest and during flight. *Insect Biochem. 16: 749-756* **enz musc flight metab**

VAANDRAGER SH VAN MARREWIJK WJA BEENAKKERS AMT 1987 Kinetic properties of glycogen phosphorylases a, ab and b from flight muscles of the locust, *Locusta migratoria. Insect Biochem. 17: 695-700* **enz musc flight metab**

VAN ANTWERPEN R BEEWILDER J VAN HEUSDEN MC VAN DER HORST DJ BEENAKKERS AMT 1990 Interaction of lipophorin with the plasma membrane of locust flight muscles. *Biol. Chem. Hoppe-Seyler 371: 159-165* **flight musc metab lip**

VAN ANTWERPEN R LINNEMANS WAM VAN DER HORST DJ BEENAKKERS AMT 1988 Immunocytochemical localization of lipoproteins in the flight muscles of the migratory locust. *Cell Tiss. Res. 252: 661-668* **musc flight lip**

VAN HEUSDEN MC VAN DER HORST DJ VAN DOORN JM BEENAKKERS AMT 1987 Partial purification of locust flight muscle lipoprotein lipase (LpL): apparent differences from mammalian LpL. *Comp. Biochem. Physiol. B 88: 523-527* **enz musc flight lip**

VAN HEUSDEN MC VAN DER HORST DJ VAN DOORN JM WES J BEENAKKERS AMT 1986 Lipoprotein lipase activity in the flight muscle of *Locusta migratoria* and its specificity for haemolymph lipoproteins. *Insect Biochem. 16: 517-523* **enz musc flight lip blood**

VAN MARLE J PIEK T LIND A VANWEERENENKRAMER J 1990 Time-related distribution of ^{3}H-glutamate in glutamergic terminal axons of the locust *Schistocerca gregaria* after uptake in vitro. *Neurosci. Res. Comm. 7: 151-157* **TS glu MN**

VANDEN BROECK J CARDOEN J NEYTS J SWINNEN K DE LOOF A 1990 Detection of substances recognized by antisera directed against vertebrate somatotropin, prolactin and placental lactogen, within the brain of *Locusta migratoria*: a comparison of immunocytochemical localization patterns. *Comp. Biochem. Physiol. A 97: 35-40* **brain pep**

VANDEN BROECK J DE LOOF A 1989 Selection of locust brain cDNA clones by antisera raised against vertebrate peptide hormones. *Gen. comp. Endocrinol. 74: 243* **hormX brain cellCult cDNA**

VANDEN BROECK J FIAS E MANNAERTS L LARIDON J DE LOOF A 1990 Selection of locust brain cDNA clones by immunological screening of an expression library. *Archs. Int. Physiol. Biochim. 98: B 48* **hormX brain cellCult cDNA**

VANDEN BROECK J VAN BRUSSEL M DE LOOF A 1990 Amplification of a locust brain cDNA sequence by means of the polymerase chain reaction (PCR). *Archs. Int. Physiol. Biochim. 98: B 49* **hormX brain cellCult cDNA**

VANHEMS E 1983 Rôle des cellules indifférenciées de la pars intercerebralis du criquet migrateur pendant la vie post-embryonnaire. *C. R. Seanc. Soc. Biol., Paris 177: 513-521* **brain NSC devel glia**

VANHEMS E 1985 An in vitro autoradiographic study of gliogenesis in the embryonic locust brain. *Dev. Brain Res. 23: 269-275* **glia brain devel**

VANHEMS E DELBOS M 1987 Differentiation of glial cells and neurite outgrowth obtained from embryonic locust central nervous system explants. *Brain Res. 411: 129-138* **glia devel cellCult**

VANHEMS E DELBOS M 1989 Ultrastructural localization of lectin-binding sites and laminin-like immunoreactivity in glial cells and neurites growing out from explant cultures of the central nervous system of embryonic locusts. *Cell Tiss. Res. 258: 429-436* **devel glia ECM cellCult**

VANHEMS E DELBOS M GIRARDIE J 1990 Insulin and neuroparsin promote neurite outgrowth in cultured locust CNS. *Eur. J. Neurosci. 2: 776-782* **pep cellCult neuropars insul**

VANHEMS E GIRARDIE J 1983 Undifferentiated cells present in the pars intercerebralis of larval and adult locust are glial precursors. Autoradiographic and ultrastructural study in vivo and in vitro. *Dev. Brain Res. 10: 177-185* **brain devel glia**

VARANKA I SVIDERSKY VL 1974 Functional characteristics of the interneurons of wind-sensitive hair receptors on the head in *Locusta migratoria.* I. Interneurons with excitatory responses. *Comp. Biochem. Physiol. A 48: 411-426* **windh SO mech IN flight**

VARANKA I SVIDERSKY VL 1974 Functional characteristics of the interneurons of wind-sensitive hair receptors on the head in *Locusta migratoria.* II. Interneurons with inhibitory responses. *Comp. Biochem. Physiol. A 48: 427-438* **windh SO mech IN flight**

VEDENINA VY ZHANTIEV RD 1990 [Recognition of acoustic signals in sympatric species of grasshoppers]. (In Russian, with English summary) *Zool. Zh. 69 (2): 36-45* **strid aud patRec**

VEENSTRA JA 1984 Immunocytochemical demonstration of a homology in peptidergic neurosecretory cells in the suboesophageal ganglion of a beetle and a locust with antisera to bovine pancreatic polypeptide, FMRFamide, vasopressin and α-MSH. *Neurosci. Lett. 48: 185-190* **pep suboes NSC vasop FMRF homol**

VEENSTRA JA 1987 Diversity in neurohaemal organs for homologous neurosecretory cells in different insect species, as demonstrated by immunocytochemistry with an antiserum to molluscan cardiac excitor peptide. *Neurosci. Lett. 73: 33-37* **pep FMRF NSC abd thor**

VEENSTRA JA YANAIHARA N 1984 Immunocytochemical localization of gastrin-releasing peptide/bombesin-like immunoreactive neurons in insects. *Histochem. 77: 113-138* **pep NSC brain anat homol**

VIEILLEMARINGE J CAILLY-LESCURE H BENSCH C GIRARDIE J 1981 Etude en histofluorescence des cellules aminergiques dans le système nerveux central du criquet migrateur. *J. Physiol. (Paris) 77: 989-995* **TS anat**

VIEILLEMARINGE J DURIS P BENSCH C GIRARDIE J 1982 Colocalization of amines and peptides in the same neurosecretory cells of locusts. *Neurosci. Lett. 31: 237-240* **pep NSC brain TS**

VIEILLEMARINGE J DURIS P GEFFARD M LE MOAL M DELAAGE M BENSCH C GIRARDIE J 1984 Immunochemical localization of dopamine in the brain of the insect *Locusta migratoria migratorioides* in comparison with the catecholamine distribution determined by the histofluoresence technique. *Cell Tiss. Res. 237: 391-394* **TS dopa anat brain retroc**

VIEILLEMARINGE J SOUAN ML GEFFARD M 1986 Immunocytochemical localization of acetylcholine receptors in locust brain. *Bull. Soc. Zool. Fr. 111: 52-53* **ACh TS brain recept**

VIEILLEMARINGE J SOUAN ML GRANDIER-VAZEILLES X GEFFARD M 1987 Immunochemical localisation of acetylcholine receptors in locust brain using auto-anti-idiotype acetylcholine antibodies. *Neurosci. Lett. 79: 59-64* **ACh TS brain recept**

VIEIRA R MARTIN F ALDEGUNDE M 1991 5-hydroxytryptamine metabolism in nervous tissue of two acridids, *Oedipoda caerulescens* and *Paracinema tricolor:* a comparative study. *Insect Biochem. 21: 545-552* TS 5HT metab brain

VILLALOBOS AR ARBAS EA 1988 Structure of motoneurons that atrophy following hindlimb autotomy in grasshoppers. *Soc. Neurosci. Abstr. 14: 688* **leg MN autot**

VINAL SC 1919 The respiratory system of the Carolina locust *Dissosteira carolina* L. *J. N.Y. entomol. Soc. 27: 19-32* **anat trach**

VINCENT JFV 1972 The dynamics of release and possible identity of bursicon in *Locusta migratoria migratorioides. J. Insect Physiol. 18: 757-780* **pep bursicon**

VIRANT-DOBERLET M HORSEMAN G LOHER W 1991 Two types of brain neuron innervate the corpora allata of the grasshopper *Gomphocerus rufus. In: Elsner N Penzlin H (eds.), Synapse-transmission-modulation: Proc. 19th Göttingen Neurobiol. Conf. Stuttgart, New York: Thieme Verlag. Abstract 328* **NSC brain retroc CAll innerv**

VISCUSO R 1971 Studio morphologico e istologico della parete epifaringea di *Eyprepocnemis plorans* (Charp.) (Orth. Acrid.). *Boll. Sedute Accad. Gioenia Sci. Nat. Catania 10: 633-655* **anat mpt**

VISCUSO R 1974 Studio comparato della lamina epifaringea di vari insetti ortotteri. *Redia 55: 129-141* **anat mpt**

VISHNEVSKAYA TM CHERKASOV AD SHURA-BURA TM 1986 {Spectral sensitivity of photoreceptors in the compound eye of the locusts (*Locusta migratoria*)}. (in Russian) *Neirofiziologiya 18: 69-76. See also Neurophysiology- Engl Tr. 18: 54-60* **ret**

VISHNEVSKAYA TM SHURA-BURA TM 1990 Spectral sensitivity of photoreceptors and spectral inputs to the neurons of the first optic ganglion in the locust (*Locusta migratoria*). *In: Gribakin FG Wiese K Popov AV*

(eds.), Sensory systems and communication in arthropods. Basel, Boston, Berlin: Birkhäuser Verlag. 106-111 **ret opt**

VISSCHER SN 1972 Peripheral ganglia in the orthopteroid insects. *Proc. Montana Acad. Sci. 32: 14* **anat head mpt**

VÖLKER A ELSNER N 1991 Laser-vibrometric measurements at the tympanic membrane of migratory locusts. *In: Elsner N Penzlin H (eds.), Synapse-transmission-modulation: Proc. 19th Göttingen Neurobiol. Conf. Stuttgart, New York: Thieme Verlag. Abstract 142* **SO tymp aud**

VON BERNHARDI R BASTIANI MJ 1990 Expression during development of a neural antigen shared by grasshopper and *Drosophila*. *Soc. Neurosci. Abstr. 16: 811* **devel**

VON HELVERSEN D 1972 Gesang des Männchens und Lautschema des Weibchens bei der Feldheuschrecke *Chorthippus biguttulus* (Orthoptera, Acrididae). *J. comp. Physiol. 81: 381-422* **strid temp patRec**

VON HELVERSEN D 1984 Parallel processing in auditory pattern recognition and directional analysis by the grasshopper *Chorthippus biguttulus* L. (Acrididae). *J. comp. Physiol. A 154: 837-846* **aud patRec**

VON HELVERSEN D 1990 Interaction of low and high frequency components in the acoustic mate recognition system of a grasshopper. *In: Elsner N Roth G (eds.), Brain - Perception - Cognition: Proc. 18th Göttingen Neurobiol. Conf., Stuttgart, New York: Thieme. p. 31* **aud behav patRec HF**

VON HELVERSEN D RHEINLAENDER J 1988 Interaural intensity and time discrimination in an unrestrained grasshopper: a tentative behavioural approach. *J. comp. Physiol. A 162: 333-340* **aud patRec**

VON HELVERSEN D VON HELVERSEN O 1975 Verhaltensgenetische Untersuchungen am akustischen Kommunikationssystem der Feldheuschrecken (Orthoptera, Acrididae). I. Der Gesang von Artbastarden zwischen *Chorthippus biguttulus* und *Ch. mollis. J. comp. Physiol. A 104: 273-299* **strid patRec**

VON HELVERSEN D VON HELVERSEN O 1975 Verhaltensgenetische Untersuchungen am akustischen Kommunikationssystem der Feldheuschrecken (Orthoptera, Acrididae). II. Das Lautschema von Artbastarden zwischen *Chorthippus biguttulus* und *Ch. mollis. J. comp. Physiol. A 104: 301-323* **strid patRec**

VON HELVERSEN D VON HELVERSEN O 1981 Korrespondenz zwischen Gesang und auslösendem Schema bei Feldheuschrecken. *Nova Acta Leopoldina N. F. 54: 449-462* **patRec strid**

VON HELVERSEN D VON HELVERSEN O 1983 Species recognition and acoustic localization in acridid grasshoppers. *In: Huber F Markl H (eds.), Neuroethology and behavioural physiology. Berlin, Heidelberg, New York: Springer Verlag. 95-107* **strid patRec**

VON HELVERSEN D VON HELVERSEN O 1985 Woran erkennen Heuschrecken den arteigenen Gesang? *In: Franck D (ed.): Verhaltensbiologie. Stuttgart, New York: Thieme Verlag. 227-231* **strid aud behav patRec**

VON HELVERSEN D VON HELVERSEN O 1990 Pattern recognition and directional analysis: routes and stations of information flow in the CNS of a grasshopper. *In: Gribakin FG Wiese K Popov AV (eds.), Sensory systems and communication in arthropods. Basel, Boston, Berlin: Birkhäuser Verlag. 209-216* **patRec aud**

VON HELVERSEN O 1979 Angeborenes Erkennen akustischer Schlüsselreize. *Verh. dt. zool. Ges. 1979: 42-59* **aud patRec**

VON HELVERSEN O 1986 Gesang und Balz bei Feldheuschrecken der *Chorthippus albomarginatus*-Gruppe (Orthoptera: Acrididae). *Zool. Jb. (Syst.) 113: 319-342* **strid behav**

VON HELVERSEN O ELSNER N 1977 The stridulatory movements of acridid grashoppers recorded with an opto-electronic device. *J. comp. Physiol. A 122: 53-64* **strid leg rhythm tech**

VON HELVERSEN O VON HELVERSEN D 1987 Innate receiver mechanisms in the acoustic communication of orthopteran insects. *In: Guthrie DM (ed.): Aims and methods in neuroethology. Manchester: Manchester University Press. 104-150* **aud patRec**

VON REKOWSKI CH HUSTERT R 1991 Sites of tactile stimuli on the body of locusts determine defensive reactions. *In: Elsner N Penzlin H (eds.), Synapse-transmission-modulation: Proc. 19th Göttingen Neurobiol. Conf. Stuttgart, New York: Thieme Verlag. Abstract 67* **mech kick SO**

VON SICARD NAE ANDERSON M CANDY DJ 1989 The effect of venom from the ant *Tetramorium caespitum* on neuromuscular transmission in the locust *Schistocerca gregaria. J. Insect Physiol. 35: 651-658* **NM tox**

VOSKRESENSKAYA AK 1947 [Functional properties of the neuromuscular apparatus in wings of insects.] (in Russian). *Sechenov J. Physiol. 33: 381-392* **flight musc thor**

VOSKRESENSKAYA AK 1950 [On "sympathetic" innervation of skeletal muscles in insects.] *Sechenov J. Physiol. 36: 176-183* **medN flight musc thor**

VOSKRESENSKAYA AK 1959 *[Functional properties of the neuro-muscular apparatus of insects.] (In Russian). Moscow & Leningrad: Nauka.* **musc thor**

VULLINGS HGB DIEDEREN JHB KONINGS PNM 1991 The innervation of the adipokinetic cells in the corpus cardiacum of *Locusta migratoria. Proc. exp. appl. Entomol., N. E. V. Amsterdam 2: 96-97* **retroc CC AKH pep innerv**

VULLINGS HGB LEGERSTEE-OOSTVEEN GG DIEDEREN JHB JANSEN WF 1989 Immunocytochemical demonstration of neuron-specific enolase in neuronal cell bodies of an insect. *Neurosci. Lett. 104: 48-52* **enz**

WADEPUHL M 1983 Control of grasshopper singing behaviour by the brain: responses to electrical stimulation. *Z. Tierpsychol. 63: 173-200* **strid brainStim**

WADEPUHL M HUBER F 1979 Elicitation of singing and courtship movements by electrical stimulation of the brain of the grasshopper. *Naturwiss. 66: 320-321* **strid brainStim behav**

WAJC E PENER MP 1971 The effect of corpora allata on the flight activity of the male african migratory locust, *Locusta migratoria migratorioides* (R. & F.). *Gen. comp. Endocrinol. 17: 327-333* **flight retroc CAll**

WALDMANN B ZARNACK W 1987 Motor activity and movements of forewings during roll manoeuvres in flying desert locusts. *In: Elsner N Creutzfeld O (eds.), New frontiers in brain research: Proc. 15th Göttingen Neurobiol. Conf. Stuttgart, New York: Thieme Verlag. Abstract 54* **flight aero**

WALDMANN B ZARNACK W 1988 Forewing movements and motor activity during roll manoeuvres in flying desert locusts. *Biol. Cybern. 59: 325-335* **flight MN aero**

WALDOW U 1970 Elektrophysiologische Untersuchungen an Feuchte-, Trocken-, und Kälterrezeptoren auf der Antenne der Wanderheuschrecke *Locusta. Z. vergl. Physiol. 69: 249-283* **ant SO hygr temp**

WALDRON I 1967 Mechanisms for the production of the motor output pattern in flying locusts. *J. exp. Biol. 47: 201-212* **flight rhythm**

WALDRON I 1967 Neural mechanisms by which controlling inputs influence motor output in the flying locust. *J. exp. Biol. 47: 213-228* **flight SO rhythm**

WALDRON I 1968 The mechanism of coupling of the locust flight oscillator to oscillatory inputs. *Z. vergl. Physiol. 57: 331-347* **flight rhythm**

WALLACE GK 1958 Some experiments on form perception in the nymphs of the desert locust, *Schistocerca gregaria* Forskål. *J. exp. Biol. 35: 765-75* **vis behav**

WALLACE GK 1959 Visual scanning in the desert locust *Schistocerca gregaria* Forskål. *J. exp. Biol. 36: 512-525* **vis jump behav**

WALLACE GK 1962 Experiments on visually controlled orientation in the desert locust, *Schistocerca gregaria* Forskål. *Anim. Behav. 10: 361-369* **vis behav**

WALOFF Z 1972 Observations on the airspeeds of freely flying locusts. *Anim. Behav. 20: 367-372* **flight behav**

WALOFF Z 1972 Orientation of flying locusts, *Schistocerca gregaria* (Forsk.), in migrating swarms. *Bull. entomol. Res. 62: 1-72* **flight behav**

WALTHER C 1980 Physiology of insect motor nerve terminals. Some effects of toxins and drugs. *In: Sherwood M (ed.): Insect neurobiology and pesticide action (Neurotox 79). London: Society of Chemical Industry Press. 153-160* **NM tox**

WALTHER C 1980 Small motor axons in orthopteran insects: a reinvestigation of the innervation of the femoral retractor unguis muscle in a stick insect and two species of locust. *J. exp. Biol. 87: 99-119* **anat innerv leg musc**

WALTHER C 1981 Synaptic terminals from an identified motoneuron in locust muscle: comparison between 1st instar larva and adult. *Neurosci. Lett. 27: 237-242* **fineStruc devel MN**

WALTHER C HODGKISS JP HENSLER K 1982 Subminiature junction potentials in excitatory neuromuscular synapses of the locust. *Neurosci. Lett. 34: 27-32* **NM**

WALTHER C RATHMAYER W 1974 The effect of *Habrobracon* venom on excitatory neuromuscular transmission in insects. *J. comp. Physiol. 89: 23-38* **NM tox**

WALTHER C SCHAFER S 1988 FMRFamide-like immunoreactivity in the metathoracic ganglion of the locust (*Schistocerca gregaria*). *Cell Tiss. Res. 253: 489-491* **FMRF pep anat thor rhythm medN**

WALTHER C SCHIEBE M 1987 Pre- and postsynaptic modulation by FMRF-NH2-like peptides in insect skeletal muscle. *In: Elsner N Creutzfeld O (eds.), New frontiers in brain research: Proc. 15th Göttingen Neurobiol. Conf. Stuttgart, New York: Thieme Verlag. Abstract 257* **FMRF pep TS mod musc MN**

WALTHER C SCHIEBE M 1987 FMRF-NH2-like factor from neurohaemal organ modulates neuromuscular transmission in the locust. *Neurosci. Lett. 77: 209* **FMRF pep NM mech medN nusc mod**

WALTHER C SCHIEBE M VOIGT KH 1984 Synaptic and non-synaptic effects of molluscan cardioexcitatory neuropeptides on locust skeletal muscle. *Neurosci. Lett. 45: 99-104* **FMRF pep NM musc mod**

WALTHER C USHERWOOD PNR 1972 Characterization of the glutamate receptors at the locust excitatory neuromuscular junction. *Verh. dt. zool. Ges. 1972: 309-312* **NM glu recept**

WALTHER C ZITLAU KE 1991 Effects of octopamine and cAMP related drugs on the resting membrane conductance in locust skeletal muscle. *In: Elsner N Penzlin H (eds.), Synapse-transmission-modulation: Proc. 19th Göttingen Neurobiol. Conf. Stuttgart, New York: Thieme Verlag. Abstract 2* **musc oct cAMP mod**

WALTHER C ZLOTKIN E RATHMAYER W 1976 Action of different toxins from the scorpion *Androctonus australis* on a locust nerve-muscle preparation. *J. Insect Physiol. 22: 1187-1194* **NM tox**

WARDAUGH KG 1978 A description of the embryonic stages of the Australian plague locust, *Chortoiocetes terminifera* (Walk.). *Acrida 7: 1-10* **devel**

WASSER G ELSNER N 1991 Wing and leg stridulation in the grasshopper *Stenobothrus rubicundus. In: Elsner N Penzlin H (eds.), Synapse-transmission-modulation: Proc. 19th Göttingen Neurobiol. Conf. Stuttgart, New York: Thieme Verlag. Abstract 144* **behav leg wing strid**

WATKINS BL BURROWS M 1989 GABA-like immunoreactivity in the suboesophageal ganglion of the locust *Schistocerca gregaria. Cell Tiss. Res. 258: 53-63* **TS GABA suboes brain neck inhib saliv**

WATKINS BL BURROWS M SIEGLER MVS 1985 The structure of locust non-spiking interneurones in relation to the anatomy of a segmental ganglion. *J. comp. Neurol. 240: 233-255* **local nonspik anat thor**

WATSON AHD 1984 The dorsal unpaired median neurons of the locust metathoracic ganglion: neuronal structure and diversity, and synapse distribution. *J. Neurocytol. 13: 303-327* **DUM anat thor**

WATSON AHD 1986 The distribution of GABA-like immunoreactivity in the thoracic nervous system of the locust *Schistocerca gregaria. Cell Tiss. Res. 246: 331-341* **TS GABA thor local**

WATSON AHD 1988 Antibodies against GABA and glutamate label neurons with morphologically distinct synaptic vesicles in the locust central nervous system. *Neuroscience 26: 33-44* **TS GABA glu fineStruc**

WATSON AHD 1989 Antibodies against GABA label processes presynaptic to sensory afferent terminals in the locust central nervous system. *In: Erber J Menzel Pflüger H-J Todt D (eds.), Neural mechanisms of behavior: Proc. 2nd Int. Congr. Neuroethology. Stuttgart & New York: Thieme Verlag. Abstract 248* **SO inhib GABA cerc mech abd**

WATSON AHD 1990 Ultrastructural evidence of GABAergic input onto cercal afferents in the locust (*Locusta migratoria*). *J. exp. Biol. 148: 509-515* **fineStruc TS GABA cerc abd SO mech inhib**

WATSON AHD BURROWS M 1981 Input and output synapses on identified motor neurones of a locust revealed by the intracellular injection of horseradish peroxidase. *Cell Tiss. Res. 215: 325-332* **MN fineStruc leg flight**

WATSON AHD BURROWS M 1982 The ultrastructure of identified locust motor neurones and their synaptic relationships. *J. comp. Neurol. 205: 383-397* **fineStruc MN leg**

WATSON AHD BURROWS M 1983 The morphology, ultrastructure and distribution of synapses on an intersegmental interneurone of the locust. *J. comp. Neurol. 214: 154-169* **IN flight fineStruc tho**

WATSON AHD BURROWS M 1985 The synaptic basis for integration of local reflexes in the lcoust. *In: Barnes WJP Gladden MH (eds.), Feedback and motor control in invertebrates and vertebrates. London, Sydney, Dover, New Hampshire: Croom Helm Ltd. 231-250* **thor local IN MN rev**

WATSON AHD BURROWS M 1985 The distribution of synapses on the two fields of neurites of spiking local interneurones in the locust. *J. comp. Neurol. 240: 219-232* **IN local fineStruc thor**

WATSON AHD BURROWS M 1987 Immunocytochemical and pharmacological evidence for GABAergic spiking local interneurones in the locust. *J. Neurosci. 7: 1741-1751* **local IN TS GABA**

WATSON AHD BURROWS M 1988 Distribution and morphology of synapses on nonspiking local interneurones in the thoracic nervous system of the locust. *J. comp. Neurol. 272: 605-616* **nonspik thor IN anat fineStruc**

WATSON AHD BURROWS M HALE JP 1985 The morphology and ultrastructure of common inhibitory motor neurones in the thoracic ganglia of the locust. *J. comp. Neurol. 239: 341-359* **MN inhib anat fineStruc**

WATSON AHD ENGLAND RCD 1991 The distribution of and interactions between GABA-immunoreactive and non-immunoreactive preocesses presynaptic to afferents from campaniform sensilla on the trochanter of the locust leg. *Cell Tiss. Res. 266: 331-342* **TS GABA fineStruc SO mech leg thor**

WATSON AHD LAURENT G 1990 GABA-like immunoreactivity in a population of locust intersegmental interneurones and their inputs. *J. comp. Neurol. 302: 761-767* **IN MN leg thor GABA inhib**

WATSON AHD PFLÜGER H-J 1988 The distribution of GABA-like reactivity in relation to ganglion structure in the abdominal nerve cord of the locust (*Schistocerca gregaria*). *Cell Tiss. Res. 249: 391-402* **GABA TS anat abd**

WATSON AHD PFLÜGER H-J 1984 The ultrastructure of prosternal sensory hair afferents within the locust central nervous system. *Neuroscience 11: 269-279* **SO mech fineStruc thor windh mech anat**

WATSON AHD PFLÜGER H-J 1989 Regional specialisation for synaptic input and output on a locust intersegmental interneurone with multiple spike-initiating zones. *J. comp. Neurol. 279: 515-527* **IN anat abd thor**

WATSON AHD STORM-MATHISEN J OTTERSEN OP 1991 GABA and glutamate-like immunoreactivity in processes presynaptic to afferents from hair plates on the proximal joints of the locust leg. *J. Neurocytol. 230: 796-809* **TS GABA fineStruc leg SO mech**

WATSON DG MIDGLEY JM BROOKS CJW 1990 Stereochemical GC-MS analysis of octopamine in a locust. *In: Frank H Holmstedt B Testa B (eds.), Chirality and biological activity. New York: Alan R. Liss Inc. 75-80* **TS oct tech**

WECKSTRÖM M 1990 Voltage-dependent potassium currents in locust photoreceptors. *In: Elsner N Roth G (eds.) Brain - Perception - Cognition: Proc. 18th Göttingen Neurobiol. Conf. Stuttgart, New York: Thieme Verlag. Abstract 203* **channel ret**

WEGENER G BEINHAUER I KLEE A NEWSHOLME EA 1987 Properties of locust muscle 6-phosphofructokinase and their importance in the regulation of glycolytic flux during prolonged flight. *J. comp. Physiol. B 157: 315-326* **metab enz flight musc**

WEGENER G MICHEL R NEWSHOLME EA 1986 Fructose 2,6-bisphosphate as a signal for changing from sugar to lipid oxidation during flight in locusts. *FEBS Lett. 201: 129-132* **metab flight**

WEI AG 1985 The neural organization of ventilation in the locust. *Dissertation, University of Oregon, USA.* **resp 2-DOG rhythm**

WEIDNER M ARBAS EA 1985 Transneuronal induction of muscle atrophy in grasshoppers. *Soc. Neurosci. Abstr. 11: 479* **autot musc**

WEIS-FOGH T 1949 An aerodynamic sense organ stimulating and regulating flight in locusts. *Nature 163: 873-874* **windh mech flight SO**

WEIS-FOGH T 1956 Tetanic force and shortening in locust flight muscle. *J. exp. Biol. 33: 668-684* **flight musc tech saline**

WEIS-FOGH T 1956 Biology and physics of locust flight. II. Flight performance of the desert locust (*Schistocerca gregaria*). *Phil. Trans. R. Soc. Lond. (Biol.) 239: 459-510* **flight**

WEIS-FOGH T 1956 Biology and physics of locust flight. IV. Notes on sensory mechanisms in locust flight. *Phil. Trans. R. Soc. Lond. (Biol.) 239: 553-584* **flight SO**

WEIS-FOGH T 1961 Power in flapping flight. *In: Ramsay JA Wigglesworth VB (eds.), The Cell and the Organism. London: Cambridge University Press. 283-300* **flight aero**

WEIS-FOGH T 1964 Diffusion in insect wing muscle, the most active tissue known. *J. exp. Biol. 41: 229-256* **flight trach resp musc**

WEIS-FOGH T 1964 Biology and physics of locust flight. VIII. Lift and metabolic rate of flying locusts. *J. exp. Biol. 41: 257-271* **flight aero metab**

WEIS-FOGH T 1973 Quick estimates of flight fitness in hovering animals, including novel mechanisms for lift production. *J. exp. Biol. 59: 169-230* **flight aero**

WEIS-FOGH T 1975 Unusual mechanisms for the generation of lift in flying animals. *Sci. Am. 233: 80-87* **flight aero**

WEIS-FOGH T JENSEN M 1956 Biology and physics of locust flight. I. Basic principles in insect flight. A critical review. *Phil. Trans. R. Soc. Lond. (Biol.) 239: 415-458* **flight aero rev**

WEISEND P 1957 Die postembryonale Entwicklung der Thorax-muskulatur bei einigen Feldheuschrecken mit besonderer Berücksichtigung der Flugmuskeln. *Z. Morphol. Ökol. Tiere 46: 529-570* **devel thor musc flight**

WEISS MJ 1981 Structural patterns in the corpora pedunculata of Orthoptera: a reduced silver analysis. *J. comp. Neurol. 203: 515-553.* **brain anat MB**

WENDLER G 1972 Einfluss erzwungener Flügelbewegungen auf das motorische Flugmuster von Heuschrecken. *Naturwiss. 5: 220* **flight SO rhythm mech**

WENDLER G 1974 The influence of proprioceptive feedback on locust flight coordination. *J. comp. Physiol. 88: 173-200* **flight SO rhythm propr rhythm**

WENDLER G 1978 The possible role of fast wing reflexes in locust flight. *Naturwiss. 65: 65* **flight SO propr**

WENDLER G 1978 Lokomotion: das Ergebnis zentral-peripherer Interaktion. *Verh. dt. zool. Ges. 1977: 80-96* **flight SO rhythm propr**

WENDLER G 1983 The locust flight system: functional aspects of sensory input and methods of investigation. *In: Nachtigall W (ed.), BIONA-report 2 "Insekten Flug"; Akad. Wiss. Mainz. Stuttgart & New York: Gustav Fischer Verlag. 113-125* **flight SO**

WENDT B HOMBERG U BUIJS RM 1990 Immunocytochemistry of dopamine in the brain of the locust *Schistocerca gregaria. In: Elsner N Roth G (eds.) Brain - Perception - Cognition: Proc. 18th Göttingen Neurobiol. Conf. Stuttgart, New York: Thieme Verlag. Abstract 317* **TS dopa brain anat**

WERMAN R MCCANN FV GRUNDFEST H 1961 Graded and all-or-none electrogenesis in the arthropod muscle. I. The effects of alkali-earth cations on the neuromuscular system of *Romalea microptera. J. gen. Physiol. 44: 979-995.* **MN musc rom**

WERNER M 1980 Vergleichende Untersuchungen an Hexokinasen des Nervengewebes von Vertebraten, Mollusken und Insekten. *Dissertation, Universität Mainz, Germany.* **metab enz**

WEVER EG 1935 A study of hearing in the sulfur-winged grasshopper (*Arphia sulphurea*). *J. comp. Psychol. 20: 17-20* **aud SO tymp**

WEVER EG VERNON JA 1957 The auditory sensitivity of the atlantic grasshopper. *Proc. nat. Acad. Sci. USA 43: 346-348* **aud SO tymp**

WEVER EG VERNON JA 1959 The auditory sensitivity of Orthoptera. *Proc. nat. Acad. Sci. USA 45: 413-419* **aud SO tymp**

WHEELER CH VAN DER HORST DJ BEENAKKERS AMT 1984 Lipolytic activity in the flight muscles of *Locusta migratoria* measured with haemolymph lipoproteins as substrates. *Insect Biochem. 14: 261-266* **enz musc flight lip**

WHEELER WM 1891 Neuroblasts in the arthropod embryo. *J. Morphol. 4: 337-343* **devel**

WHEELER WM 1893 A contribution to insect embryology. *J. Morphol. 8: 1-160* **devel**

WHIM MD EVANS PD 1988 Octopaminergic modulation of locust flight muscle. *In: Boulton AA Juorio AV Downer RGH (eds.): Trace amines: Comparative, neurobiological and clinical studies. Clifton, New Jersey: Humana Press. 75-80* **TS oct musc flight mod**

WHIM MD EVANS PD 1988 Octopaminergic modulation of flight muscle in the locust. *J. exp. Biol. 134: 247-266* **TS oct musc flight mod**

WHIM MD EVANS PD 1989 Age-dependence of octopaminergic modulation of flight muscle in the locust. *J. comp. Physiol. A 165: 125-137* **musc TS oct mod flight devel**

WHIM DM EVANS PD 1991 The role of cyclic AMP in the octopaminergic modulation of flight muscle in the locust. *J. exp. Biol. 161: 423-438* **flight musc mod oct cAMP**

WHITE PR CHAPMAN RF 1990 Olfactory sensitivity of gomphocerine grasshoppers to the odours of host and non-host plants. *Entomol. exp. appl. 55: 205-212* **feed behav chem SO ant**

WHITE PR CHAPMAN RF 1990 Tarsal chemoreception in the polyphagous grasshopper *Schistocerca americana*: behavioural assays, sensilla distributions and electrophysiology. *Physiol. Entomol. 15: 105-121* **feed behav chem SO leg**

WHITE PR CHAPMAN RF ASCOLI-CHRISTENSEN A 1990 Interactions between two neurons in contact chemosensilla of the grasshopper, *Schistocerca americana. J. comp. Physiol. A 167: 431-436* **SO chem leg inhib**

WHITINGTON PM 1985 Functional connections with foreign muscles made by a target-deprived insect motoneuron. *Dev. Biol. 107: 537-540* **devel MN thor leg**

WHITINGTON PM 1989 The early development of motor axon pathways in the locust embryo: the establishment of the segmental nerves in the thoracic ganglia. *Development 105: 715-721* **devel MN thor Inhib leg**

WHITINGTON PM BATE M SEIFERT E RIDGE K GOODMAN CS 1982 Survival and differentiation of identified embryonic neurons in the absence of their target muscles. *Science 215: 973-975* **devel MN thor DUM leg**

WHITINGTON PM MYERS CM BALL EE 1988 Pathfinding by motor axons in the locust embryo. *Proc. 18th. Int. Congr. Entomol., Vancouver, Canada: 76* **devel MN**

WHITINGTON PM SEIFERT E 1981 Identified neurons in an insect embryo: the pattern of neurons innervating the metathoracic leg of the locust. *J. comp. Neurol. 200: 203-212* **devel MN thor leg**

WHITINGTON PM SEIFERT E 1982 Axon growth from limb motoneurons in the locust embryo: the effect of target limb removal on the path taken out of the central nervous system. *Dev. Biol. 93: 206-215* **devel MN thor leg**

WHITINGTON PM SEIFERT E 1984 Axon growth from limb motoneurons in the locust embryo: the effect of target limb removal on the pattern of axon branching in the periphery. *Dev. Biol. 93: 438-449* **devel MN thor leg**

WHITMAN DW 1982 Grasshopper sexual pheromone: a component of the defensive secretion in *Taeniopoda eques. Physiol. Entomol. 7:111-115* **pher rom**

WHITMAN DW 1990 Grasshopper chemical communication. *In: Chapman RF Joern A (eds.), Biology of grasshoppers. New York: J. Wiley & Sons Inc. 357-392* **pher**

WHITTON PW NICHOLSON RA STRANG RHC 1987 Anaerobic aminoacid metabolism in the locust nervous system. *Trans. Biochem. Soc. 14: 609-610* **metab**

WHITTON PW NICHOLSON RA STRANG RHC 1988 The effect of taurine on neurotransmitter release from insect synaptosomes. *J. Neurochem. 51: 1356-1361* **TS taur synaptosome**

WHITTON PW STRANG RHC NICHOLSON RA 1987 The distribution of taurine in insect tissues. *Insect Biochem. 17: 573-577* **TS taur**

WHITTON PW STRANG RHC NICHOLSON RA 1988 The effect of taurine on the efflux of calcium from locust mitochondria. *Insect Biochem. 18: 359-362* **TS taur Ca2**

WHITTON PW STRANG RHC NICHOLSON RA 1988 The effect of taurine on calcium accumulation in resting and depolarised insect synaptosomes. *J. Neurochem. 50: 1743-1746* **TS taur Ca2 synaptosome**

WIESEND P 1957 Die postembryonale Entwicklung der Thoraxmusculatur bei einigen Feldheuschrecken mit besonderer Berücksichtigung der Flugmuskeln. *Z. Morphol. Oekol. Tiere 46: 259-270* **musc thor anat devel flight**

WILKIN PJ 1985 Aerodynamics. *In: Kerkut GA Gilbert LI (eds.), Comprehensive insect physiology, biochemistry and pharmacology. Vol. 8. London: Pergamon Press. 553-570* **flight behav aero**

WILLIAMS DS 1983 Changes of photoreceptor performance associated with the daily turn-over of photoreceptor membrane in locusts. *J. comp. Physiol. A 150: 509-520* **ret**

WILLIAMS JLD 1972 Some observations on the neuronal organization of the supraoesophageal ganglion in *Schistocerca gregaria* Forskål with particular reference to the central complex. *Dissertation, University of Wales, UK.* **anat brain DN CB**

WILLIAMS JLD 1975 Anatomical studies of the insect central nervous system: a ground-plan of the mid-brain and an introduction to the central complex in the locust, *Schistocerca gregaria* (Orthoptera). *J. Zool. (Lond.) 176: 67-86* **anat brain DN CB**

WILLIAMSON R BURNS MD 1978 Multiterminal receptors in the locust mesothoracic leg. *J. Insect Physiol. 24: 661-666* **SO leg thor propr**

WILLIAMSON R BURNS MD 1982 Large neurones in locust neck connectives. I. Responses to sensory inputs. *J. comp. Physiol. A 147: 379-388* **anat DN**

WILSON DM 1961 The central nervous control of flight in a locust. *J. exp. Biol. 38: 471-490* **flight rhythm**

WILSON DM 1962 Bifunctional muscles in the thorax of grasshoppers. *J. exp. Biol. 39: 669-677* **flight musc anat thor**

WILSON DM 1963 Motor and sensory mechanisms of lift control in locust flight. *Proc. 16th. Int. Congr. Zool., Washington DC, 1963. 10.* **flight SO**

WILSON DM 1964 The origin of the flight motor command in grasshoppers. *In: Reiss RF (ed.), Neural Theory and Modelling. Stanford, California: Stanford University Press. 331-345* **flight rhythm MN**

WILSON DM 1964 Relative refractoriness and patterned discharge of locust flight motor neurons. *J. exp. Biol. 41: 191-205* **flight MN rhythm**

WILSON DM 1966 Central nervous mechanisms for the generation of rhythmic behaviour in arthropods. *In: Hughes GM (ed.) Nervous and hormonal mechanisms of integration, Symp. Soc. Exp. Biol. 20. Cambridge: Cambridge University Press. 199-228* **flight rhythm rev**

WILSON DM 1968 The nervous control of flight and related behaviour. *Adv. Insect Physiol. 5: 289-338* **flight rev anat rhythm**

WILSON DM 1968 Inherent assymetry and reflex modulation of the locust flight motor pattern. *J. exp. Biol. 48: 631-641* **flight OM vis**

WILSON DM 1968 The flight control system of the locust. *Sci. Am. 218: 83-90* **flight rev**

WILSON DM 1972 Stabilising mechanisms in insect flight. *In:* Hemming CF Taylor THC *(eds.), Proc. Internat. Study Conf. Current and future problems in Acridology, 1970. London: HMSO. 47-52* **flight rev OM vis**

WILSON DM GETTRUP E 1963 A stretch reflex controlling wingbeat frequency in grasshoppers. *J. exp. Biol. 40: 171-186* **flight SO SR propr rhythm**

WILSON DM WEIS-FOGH T 1962 Patterned activity of coordinated motor units, studied in flying locusts. *J. exp. Biol. 39: 643-668* **flight rhythm musc**

WILSON DM WYMAN RJ 1965 Motor output patterns during random and rhythmic stimulation of locust thoracic ganglia. *Biophys. J. 5: 121-143* **flight rhythm**

WILSON JA 1979 The structure and function of serially homologous leg motor neurons in the locust. I. Anatomy. *J. Neurobiol. 10: 41-45* **MN leg anat homol**

WILSON JA 1979 The structure and function of serially homologous leg motor neurons in the locust. II. Physiology. *J. Neurobiol. 10: 153-167* **anat MN leg var homol**

WILSON JA 1981 Unique, identifiable local nonspiking interneurons in the locust mesothoracic ganglion. *J. Neurobiol. 12: 353-366* **nonspik IN thor anat**

WILSON JA HOYLE G 1978 Serially homologous neurones as concommitants of functional specialisation. *Nature 274: 377-379* **MN leg anat homol**

WILSON JA PHILLIPS CE 1982 Locust local nonspiking interneurons which tonically drive antagonistic motor neurons: physiology, morphology and ultrastructure. *J. comp. Neurol. 204: 21-31* **nonspik IN leg MN anat**

WILSON JA PHILLIPS CE 1983 Pre-motor local nonspiking interneurons: a review. *Prog. Neurobiol. 20: 89-107* **thor leg IN nonspik rev**

WILSON JA PHILLIPS CE ADAMS ME HUBER F 1982 Structural comparison of a homologous neuron in gryllid and acridid insects. *J. Neurobiol. 13: 459-467* **MN anat thor homol**

WILSON M 1975 Angular sensitivity of light and dark adapted locust retinula cells. *J. comp. Physiol. A 97: 323-328* **ret**

WILSON M 1975 Autonomous pigment movement in the radial pupil of locust ocelli. *Nature 258: 603-604* **oc**

WILSON M 1978 The functional organization of locust ocelli. *J. comp. Physiol. A 124: 297-316* **vis oc IN brain**

WILSON M 1978 Generation of graded potential signals in the second order cells of locust ocellus. *J. comp. Physiol. A 124: 317-331* **oc IN brain**

WILSON M 1978 The origin and properties of discrete hyperpolarising potentials in the second order cells of locust ocellus. *J. comp. Physiol. A 128: 347-358* **oc IN brain**

WILSON M GARRARD P McGINNESS S 1978 The unit structure of the locust compound eye. *Cell Tiss. Res. 195: 205-226* **opt ret anat**

WINSTANLEY C BLANEY WM 1978 Chemosensory systems of locusts in relation to feeding. *Entomol. exp. appl. 24: 750-758* **SO feed mpt chem**

WINDERER H SMOLA U 1982 Contact chemoreceptors among wind-sensitive hairs of *Locusta migratoria* (Orthoptera: Acrididae). *Int. J. Insect Morphol. Embryol. 11: 147-160.* **SO mech chem head**

WINTER DL 1967 Intracellular responses from the grasshopper eye. *Nature 213: 607-608* **ret**

WOLF H 1985 Monitoring the activity of an auditory interneuron in a free-moving grasshopper. *In: Kalmring K Elsner N (eds.) Acoustic and vibrational communication in insects. Hamburg & New York: Paul Parey Verlag. 51-60* **IN aud thor**

WOLF H 1986 Response patterns of two auditory interneurons in a freely moving grasshopper (*Chorthippus biguttulus* L.). I. Response properties in the intact animal. *J. comp. Physiol. A 158: 689-696* **aud IN thor**

WOLF H 1986 Response patterns of two auditory interneurons in a freely moving grasshopper (*Chorthippus biguttulus* L.). II. Representation of stimulus parameters relevant to behaviour. *J. comp. Physiol. A 158: 697-703* **aud IN anat thor**

WOLF H 1989 The forewing steering muscle of the locust: its function and inhibitory innervation. *In: Elsner N Singer W (eds.), Dynamics and plasticity in neuronal systems: Proc. 17th. Göttingen Neurobiology Conference. Stuttgart, New York: Thieme Verlag. Abstract 110* **flight MN inhib musc**

WOLF H 1989 Intracellular recording of neural activity in semi-intact walking locusts. *In: Erber J Menzel R Pflüger H-J Todt D (eds.), Neural mechanisms of behavior: Proc. 2nd Internat. Congr. Neuroethology. Stuttgart & New York: Thieme Verlag. Abstract 7* **MN inhib walk tech**

WOLF H 1990 On the function of a locust flight steering muscle and its inhibitory innervation. *J. exp. Biol. 150: 55-80* **flight musc inhib**

WOLF H 1990 Activity patterns of inhibitory motoneurones and their impact on leg movement in tethered walking locusts. *J. exp. Biol. 152: 281-304* **leg MN inhib walk**

WOLF H 1991 Reflex modulation in locusts walking on a treadwheel - intracellular recordings from motoneurons. *In: Elsner N Penzlin H (eds.), Synapse-transmission-modulation: Proc. 19th Göttingen Neurobiol. Conf. Stuttgart, New York: Thieme Verlag. Abstract 54* **MN walk reflex**

WOLF H 1991 Sensory feedback in locust flight patterning. *In: Armstrong DM Bush BMH (eds.): Locomotor neural control in arthropods and vertebrates. Manchester: Manchester University Press. 134-148* **flight SO rhythm propr**

WOLF H PEARSON KG 1987 Intracellular recordings in tethered flying locusts: studying the role of the CNS and periphery in flight pattern generation. *In: Elsner N Creutzfeld O (eds.), New frontiers in brain research: Proc. 15th Göttingen Neurobiol. Conf. Stuttgart, New York: Thieme Verlag. Abstract 7* **flight MN SR rhythm propr**

WOLF H PEARSON KG 1987 On the question of the segmental organization of the locust flight oscillator. *In: Elsner N Creutzfeld O (eds.), New frontiers in brain research: Proc. 15th Göttingen Neurobiol. Conf. Stuttgart, New York: Thieme Verlag. Abstract 55* **flight rhythm**

WOLF H PEARSON KG 1987 Comparison of motor patterns in the intact and deafferented flight system of the locust. II. Intracellular recordings from flight motoneurons. *J. comp. Physiol. A 160: 269-279* **flight MN rhythm propr thor**

WOLF H PEARSON KG 1987 Flight motor pattterns recorded in surgically isolated sections of the ventral nerve cord of *Locusta migratoria*. *J. comp. Physiol. A 161: 103-114* **flight MN rhythm thor**

WOLF H PEARSON KG 1987 Intracellular recordings from interneurons and motoneurons in intact flying locusts. *J. Neurosci. Meth. 21: 345-354* **tech flight IN MN thor**

WOLF H PEARSON KG 1988 Locust flight pattern generation: activity patterns of a sensory input to identified flight interneurons. *In: Elsner N Barth FG (eds.), Sense organs: interfaces between environment and behaviour: Proc.16th. Göttingen Neurobiol. Conf. Stuttgart, New York: Thieme Verlag. Abstract 126* **flight IN SO rhythm**

WOLF H PEARSON KG 1988 Proprioceptive input patterns elevator activity in the locust flight system. *J. Neurophysiol. 59: 1831-1853* **flight IN rhythm SO propr**

WOLF H PEARSON KG 1989 Comparison of motor patterns in the intact and deafferented flight system of the locust. III. Patterns of interneuronal activity. *J. comp. Physiol. A 165: 61-74* **flight MN rhythm IN SR teg**

WOLF H RONACHER B REICHERT H 1988 Patterned synaptic drive to locust flight motoneurons after hemisection of thoracic ganglia. *J. comp. Physiol. A 163: 761-769* **flight thor MN IN rhythm**

WOLF H SCHMIDT J 1990 Sensory input to locust common inhibitory motoneurons during walking. *Verh. dt. zool. Ges. 83: 437.* **inhib walk SO**

WOLF H VON HELVERSEN O 1986 "Switching-off" of an auditory interneuron during stridulation in the acridid grasshopper *Chorthippus biguttulus. J. comp. Physiol. A 158: 861-871* **aud IN facnonresp**

WOOD DW 1963 The sodium and potassium composition of some insect skeletal muscle fibres in relation to their membrane potentials. *Comp. Biochem. Physiol. 9: 151-159* **musc**

WOOD SJ OSBORNE RH CATTELL KJ BANNER SE 1990 FMRFamide and related peptides modulate the actions of 5-hydroxytryptamine and proctolin on the foregut of the locust, *Schistocerca gegaria. Biochem. Soc. Trans. 18: 384-385* **gut pep 5HT proct FMRF mod**

WOOD SJ OSBORNE RH CATTELL KJ BANNER SE 1990 Nicotine relaxes the isolated foregut of the locust *Schistocerca gregaria* by causing release of endogenous octopamine. *J. Physiol. (Lond.) 425: 47P* **gut oct TS ACh mod**

WOODROW DF 1965 The responses of the african migratory locust *Locusta migratoria migratorioides* R. & F. to the chemical composition of the soil at oviposition. *Anim. Behav. 13: 348-356* **SO ovip genit chem**

WOODROW DF 1965 Laboratory analysis of oviposition behaviour in the red locust. *Bull. entomol. Res. 55: 733-745* **ovip genit behav**

WOOLLACOTT M HOYLE G 1976 Membrane resistance changes associated with single, identified neuron learning. *Soc. Neurosci. Abstr. 6: 471* **MN learn thor**

WOOLLACOTT M HOYLE G 1977 Neural events underlying learning in insects: changes in pacemaker. *Proc. R. Soc. Lond. (Biol.) 195: 395-415* **learn thor MN**

WORDEN MK WITTEN JL O'SHEA M 1985 Proctolin is a co-transmitter for the SETi motoneuron. *Soc. Neurosci. Abstr. 11: 327* **pep proct musc TS leg**

WORTMANN M ZARNACK W 1987 The influence of several parameters of wing movement on the lift of *Schistocerca gregaria* during descending, horizontal or ascending flight. *In: Elsner N Creutzfeld O (eds.), New frontiers in brain research: Proc. 15th Göttingen Neurobiol. Conf. Stuttgart, New York: Thieme Verlag. Abstract 53* **flight aero**

WORTMANN M ZARNACK W 1991 Quick measurements of lift, thrust and torque during flight manoeuvres of locusts. *In: Elsner N Penzlin H (eds.), Synapse-transmission-modulation: Proc. 19th Göttingen Neurobiol. Conf. Stuttgart, New York: Thieme Verlag. Abstract 85* **flight aero tech**

WULFF VJ FRY WJ LINDE FA 1955 Retinal action-potential theory and experimental results from grasshopper eyes. *J. cell. comp. Physiol. 45: 247-263* **ret**

WULFF VJ JAHN TL 1948 Retinal-nerve interval in the grasshopper. *J. Neurophysiol. 11: 117-123* **ret**

WÜRDEN S HOMBERG U DIRCKSEN H RAO KR 1990 Comparative morphology of PDH-immunoreactive neurons in the brain of orthopteroid insects. *In: Elsner N Roth G (eds.), Brain - Perception - Cognition: Proc. 18th Göttingen Neurobiol. Conf., Stuttgart, New York: Thieme. p. 316* **NSC brain pep**

YAMASHIKI N KAWAMURA K 1990 Formation of the contractile ring in dividing neuroblasts of *Chortophaga. Ann. N. Y. Acad. Sci. 582: 314-317* **devel**

YANG Q-Z BURROWS M 1983 The identification of motor neurones innervating an abdominal ventilatory muscle in the locust. *J. exp. Biol. 107: 115-128* **MN anat abd resp inhib musc**

YINON U SHULOV A TSVILICH R 1971 Audition in the desert locust: behavioural and neurophysiological studies. *J. exp. Biol. 55: 713-726* **aud flight SO tymp**

YOUNG AJ 1971 Studies on the acoustic behaviour of certain Orthoptera. *Anim. Behav. 19: 727-744* **strid behav**

YUDINA IA KNYAZEVA NI SVIDERSKY VL 1969 [Reflex mechanisms of initiation and inhibition of flight in *Locusta migratoria* L.] (in Russian). *Tr. Vses. Entomol. Obsch. 53: 369-373* **SO flight leg**

ZACHWILICHOWSKI J 1934 Über die Innervierung und die Sinnesorgane der Flügel der Feldheuschrecke *Stauroderus biguttulus* (L.). *Bull. int. acad. Cracovie B, 1934: 187-196* **SO mech thor flight innerv wing**

ZARETSKY M 1982 Quantitative measurements of a centrally generated saccadic suppression in a locust movement detector neurone. *J. Physiol. (Lond.) 328: 521-533* **DN vis inhib**

ZARETSKY M ROWELL CHF 1979 Saccadic suppression by corollary discharge in the locust. *Nature 280: 583-585* **DN vis inhib**

ZARNACK W 1969 Kinematik der Flügelbewegungen bei *Locusta migratoria* L. *Dissertation, Universität München, Germany.* **flight aero**

ZARNACK W 1970 Randwertprobleme des Insektenfluges in Hinblick auf die Berechnung und Messung aerodynamischer Kräfte. *Verh. dt. zool. Ges. 1970: 132-136* **flight aero**

ZARNACK W 1972 Flugbiophysik der Wanderheuschrecke (*Locusta migratoria* L.). I. Die Bewegung der Vorderflügel. *J. comp. Physiol. 78: 356-395* **flight aero**

ZARNACK W 1975 Aerodynamic forces and their calculation in insect flight. *In: Wu TY-T Brokaw CJ Brennan C (eds.) Symposium on swimming and flying in nature. Vol II. New York: Plenum Press. 797-801* **flight aero**

ZARNACK W 1977 A high-speed stereophotograph measuring system for flying animals. *Fortschr. Zool. 24: 317-319* **flight aero tech**

ZARNACK W 1978 A transducer recording continuously 3-dimensional rotations of biological objects. *J. comp. Physiol. A 126: 161-168* **tech flight aero**

ZARNACK W 1978 Locust flight control. On-line measurements of phase shifting in forewing movements. *Naturwiss. 65: 64* **flight aero OM**

ZARNACK W 1982 Kinematische, aerodynamische und neurophysiologisch-funktions-morphologische Untersuchungen des Heuschreckenflugs. *Habilitationsschrift, Universität Göttingen, Germany.* **flight aero**

ZARNACK W 1983 Untersuchungen zum Flug von Wanderheuschrecken: die Bewegungen, räumlichen Lagebeziehungen sowie Formen und Profile von Vorder- und Hinterflügeln. *In: Nachtigall W (ed.), BIONA-report 1 "Insekten Flug"; Akad. Wiss. Mainz. Stuttgart & New York: Gustav Fischer Verlag. 79-102* **flight aero**

ZARNACK W 1988 The effect of forewing depressor activity on wing movement during locust flight. *Biol. Cybern. 59: 55-70* **flight MN musc aero**

ZARNACK W 1988 On the latency of flight manoeuvres in locusts. *In: Elsner N Barth FG (eds.), Sense organs: interfaces between environment and behaviour: Proc.16th. Göttingen Neurobiol. Conf. Stuttgart, New York: Thieme Verlag. Abstract 122* **flight**

ZARNACK W MÖHL B 1977 Activity of the direct downstroke flight muscles of *Locusta migratoria* (L.) during steering behaviour in flight. I. Patterns of time shift. *J. comp. Physiol. A 118: 215-233* **flight**

ZARNACK W REUSE G SCHWENNE T 1991 Flight manoeuvres of locusts. *In: Alt W Hoffmann G (eds.): Biological motion. Berlin, Heidelberg, New York: Springer Verlag. 1-10* **flight aero**

ZARNACK W WORTMANN M 1989 On the so-called constant lift reaction of migratory locusts. *J. exp. Biol. 147: 111-123* **flight aero**

ZHANTIEV RD 1969 [The ultrastructure of mechanoreceptor sensillia in insects.] (in Russian). *J. Biol., Akad. Nauk USSR 30: 224-230* **aud SO tymp fineStruc**

ZHANTIEV RE KALINKINA IN 1977 [Response to sound stimuli of orthopteran descending neurons in the abdominal part of the ventral nerve cord]. (in Russian) *Nauchn. doklady visshey skoly. Biol. Nauki: 66-71* **aud IN abd**

ZIEGLER R RYAN RO ARBAS EA LAW JH 1988 The adipokinetic response of flightless grasshoppers (*Barytettix psolus*): functional components, defective response. *Archs. Insect Biochem. Physiol. 9: 255-268* **pep AKH flight**

ZILL SN 1985 Plasticity and proprioception in insects. I. Responses and cellular properties of individual receptors of the locust metathoracic femoral chordotonal organ. *J. exp. Biol. 116: 435-461* **SO propr FCO leg thor labil**

ZILL SN 1985 Plasticity and proprioception in insects. II. Modes of reflex action of the locust metathoracic femoral chordotonal organ. *J. exp. Biol. 116: 463-480* **SO propr FCO leg thor labil**

ZILL SN 1987 Selective mechanical stimulation of an identified proprioceptor in freely moving locusts: role of resistance reflexes in active posture. *Brain Res. 417: 195* **SO propr FCO leg thor**

ZILL SN FORMAN RR 1983 Proprioceptive reflexes change when an insect assumes an active, learned posture. *J. exp. Biol. 107: 385-390* **propr learn thor labil**

ZILL SN FRAZIER SF 1990 Responses of locusts in a paradigm which tests postural load compensatory reactions. *Brain Res. 535:1-8* **SO FCO propr behav walk leg thor**

ZILL SN FRAZIER SF 1990 Range fractionation in responses of spiking interneurons that detect joint movement in locusts. *Soc. Neurosci. Abstr. 16: 117* **mech propr leg IN thor**

ZILL SN FRAZIER SF 1991 Proprioceptive inputs and inherent muscle tensions contribute to load compensatory reactions in locusts. *Soc. Neurosci. Abstr. 17: 638* **reflex propr leg musc**

ZILL SN JEPSON-INNES K 1988 Evolutionary adaptation of a reflex system: sensory hysteresis counters muscle "catch" tension. *J. comp. Physiol. A 164: 43-48* **FCO SO propr MN leg thor**

ZILL SN JEPSON-INNES K 1990 Functions of a proprioceptive sense organ in freely moving insects - characteristics of reflexes elicited by stimulation of the locust metathoracic femoral chordotonal organ. *Brain Res. 523: 211-218* **FCO SO propr behav walk leg thor**

ZILL SN UNDERWOOD MA ROWLEY JC MORAN DT 1980 A somatotopic organization of groups of afferents in insect peripheral nerves. *Brain Res. 198: 253-269* **MN SO anat thor**

ZIMMERMANN R MÖHL B 1990 Effect of atropine on the stability of flight control in *Locusta migratoria* L. *In: Elsner N Roth G (eds.) Brain - Perception - Cognition: Proc. 18th Göttingen Neurobiol. Conf. Stuttgart, New York: Thieme Verlag. Abstract 76* **flight TS ACh**

ZINN K McALLISTER L GOODMAN CS 1988 Sequence analysis and neuronal expression of fasciclin I in grasshopper and *Drosophila*. *Cell 53: 577-587* **devel**